Contributions To Phenomenology

In Cooperation with The Center
for Advanced Research in Phenomenology

Volume 95

Scope

The purpose of the series is to serve as a vehicle for the pursuit of phenomenological research across a broad spectrum, including cross-over developments with other fields of inquiry such as the social sciences and cognitive science. Since its establishment in 1987, *Contributions to Phenomenology* has published more than 80 titles on diverse themes of phenomenological philosophy. In addition to welcoming monographs and collections of papers in established areas of scholarship, the series encourages original work in phenomenology. The breadth and depth of the Series reflects the rich and varied significance of phenomenological thinking for seminal questions of human inquiry as well as the increasingly international reach of phenomenological research.

The series is published in cooperation with The Center for Advanced Research in Phenomenology.

More information about this series at http://www.springer.com/series/5811

Thomas Hünefeldt • Annika Schlitte
Editors

Situatedness and Place

Multidisciplinary Perspectives on the Spatio-temporal Contingency of Human Life

Springer

Editors
Thomas Hünefeldt
Sapienza University of Rome
Rome, Italy

Annika Schlitte
Johannes Gutenberg University
Mainz, Germany

ISSN 0923-9545 ISSN 2215-1915 (electronic)
Contributions To Phenomenology
ISBN 978-3-319-92936-1 ISBN 978-3-319-92937-8 (eBook)
https://doi.org/10.1007/978-3-319-92937-8

Library of Congress Control Number: 2018954740

This Springer imprint is published by the registered company Springer Nature Switzerland AG
The registered company address is: Gewerbestrasse 11, 6330 Cham, Switzerland

Preface

This volume assembles researches that have been presented in the context of a series of events organized by the interdisciplinary "Graduate Research Training Group 'Philosophy of Place'" (Graduiertenkolleg "Philosophie des Ortes") at the Catholic University of Eichstätt-Ingolstadt. In particular, it contains works by international guests invited to a lecture series entitled "Situated Cognition and the Philosophy of Place," which took place in Eichstätt in spring 2015, or to a symposium with the same title, which took place in the context of the "6th International Conference on Spatial Cognition: Space and Situated Cognition (ICSC 2015)" in Rome. Both the lecture series and the symposium were aimed at exploring the possibilities and conditions of an interdisciplinary dialogue between two heterogeneous research movements which both emphasize the spatiotemporal contingency of human life and which both are influenced and inspired by the phenomenological tradition, but which nevertheless largely ignore or even disregard each other: the "situated cognition" movement in pedagogy, psychology, and the cognitive sciences, on the one hand, and the thought on "place" in philosophy and the social and cultural sciences, on the other.

Given this origin, the contributions to this volume are quite heterogeneous both in their methods and in their topics. They do not pretend to represent the whole breadth of the research tradition of which they are part, but present research in progress on important topics within that tradition, with an either critical or empathetic eye on motives belonging to the respective other research tradition. The first three contributions, by Edward Casey, Jeff Malpas, and David Seamon, address fundamental questions of a philosophy of place. They suggest that contemporary thought on place is far from being homogeneous, in particular as regards the role attributed to phenomenology. The following three contributions, by Shaun Gallagher, William Clancey, and Thomas Hünefeldt, focus in rather different ways on the situatedness of mental processes. Next, the contributions by Annika Schlitte and Tsutomu Ben Yagi analyze the work of two important twentieth century philosophers: Helmuth Plessner and Hans-Georg Gadamer, with respect to questions concerning situatedness and place. Last but not least, the contributions by Tobias Holischka and Dylan

Trigg deal with particular topics closely related to situatedness and place: agoraphobia and virtual reality.

We wish to thank all those who have rendered possible this volume and the context in which it originated. First, we wish of course to thank the contributors to this volume: it has been a great honor that they accepted our invitation, and a great pleasure to interact with them. Furthermore, we thank the Catholic University of Eichstätt-Ingolstadt and the "Cassianeum Stiftung," which generously sponsored the "Graduate Research Training Group 'Philosophy of Place'" and the events organized in its context. Finally, our special thanks go to Daniel Romić at the Research Office of the Catholic University of Eichstätt-Ingolstadt, who empathically supported our work throughout the entire process.

Rome, Italy — Thomas Hünefeldt
Mainz, Germany — Annika Schlitte

Contents

About the Editors and Authors

Edward S. Casey works in aesthetics, philosophy of space and time, ethics, perception, psychoanalytic theory, and environmental philosophy. His published books include *Imagining: A Phenomenological Study* (Indiana University Press, 2nd ed. 2000); *Remembering: A Phenomenological Study* (Indiana University Press, 2nd ed. 2000); *Getting Back into Place* (Indiana University Press, 2nd ed. 2009); and *The Fate of Place* (University of California Press, 2nd ed. 2013). More recently, he has undertaken studies in peri-phenomenology, publishing *The World at a Glance* (Indiana University Press, 2007) – a new look at visual perception – and *The World on Edge* (Indiana University Press, 2017), an examination of the precarious aspects of life on earth today.

William J. Clancey (Computer Science Ph.D. Stanford; Mathematical Sciences B.A. Rice) is a senior research scientist at the Florida Institute for Human and Machine Cognition. His research has related cognitive and social science in the study of work practices and the design of agent systems at the Institute for Research on Learning and as Chief Scientist of Human-Centered Computing, NASA Ames. He is a Fellow of Assoc. of Psychological Science, Assoc. Advancement of AI, National Academy of Inventors, and American College of Medical Informatics.

Shaun Gallagher is the Lillian and Morrie Moss Professor of Excellence in Philosophy at the University of Memphis (USA) and Professorial Fellow at the University of Wollongong (Australia). Professor Gallagher holds the Humboldt Foundation's Anneliese Maier Research Award [Anneliese Maier-Forschungspreis] (2012–2018). His new book *Enactivist Interventions* was published by Oxford University Press in 2017.

Tobias Holischka Ph.D., is a research assistant at the Chair of Philosophy at the Catholic University of Eichstätt-Ingolstadt and an associate member of the interdisciplinary research group "Philosophy of Place." His Ph.D. project "CyberPlaces" was on the philosophy of virtual places. His research interests focus on philosophy of technology and the relation of virtuality and reality.

Thomas Hünefeldt (Ph.D. in Philosophy, Tübingen, and in Cognitive Psychology, Sapienza University of Rome) is research fellow at the Sapienza University of Rome and Managing Editor of "Cognitive Processing – International Quarterly of Cognitive Science." His research interests focus on questions concerning intersubjectivity, both from a philosophical and from an empirical perspective.

Jeff Malpas is Distinguished Professor at the University of Tasmania in Hobart. He is the editor of *The Intelligence of Place* (Bloomsbury, 2015), and his *Place and Experience* (Cambridge, 1999) is soon to be republished in a new expanded edition by Routledge. He has also written extensively on many other topics in the history of philosophy, phenomenology, hermeneutics, philosophy of language, ethics, and political philosophy.

Annika Schlitte is Junior Professor for philosophy with a special focus on the relation to the social and cultural sciences at Johannes Gutenberg University of Mainz. She wrote her dissertation on Georg Simmel's *Philosophy of Money* (*Die Macht des Geldes und die Symbolik der Kultur,*" Fink, 2012) and was speaker of the Interdisciplinary Research Academy "Philosophy of Place" at the Catholic University of Eichstätt-Ingolstadt. Her research interests include philosophy of culture, phenomenology, and hermeneutics.

David Seamon is a Professor of Architecture at Kansas State University in Manhattan, Kansas, USA. Trained in geography and environment-behavior research, he is interested in a phenomenological approach to place, architecture, and environmental design as place making. His most recent book is *Life Takes Place: Phenomenology, Lifeworlds, and Place Making* (Routledge, 2018). He edits the *Environmental and Architectural Phenomenology Newsletter*, which in 2014 celebrated 25 years of publication.

Dylan Trigg is FWF Lise Meitner Senior Fellow at the University of Vienna, Department of Philosophy. His research specialism concerns phenomenology, aesthetics, and theories of space and place. He is the author of several books, most recently *Topophobia* (2016), *The Thing* (2014), and with Dorothée Legrand, *Unconsciousness Between Phenomenology and Psychoanalysis* (2017).

Tsutomu Ben Yagi is a Ph.D. candidate at Catholic University Eichstätt-Ingolstadt. His research centers on the questions of the mother tongue and the place of language in Gadamer's hermeneutics.

Introduction: Situatedness and Place

Thomas Hünefeldt and Annika Schlitte

Over the last two or three decades, the spatio-temporal contingency of human life has become an important topic of research in a broad range of different disciplines including the social sciences, the cultural sciences, the cognitive sciences, and philosophy. Significantly, however, this research topic is referred to in quite different ways: While some researchers refer to it in terms of the "situatedness" of human experience and action, others refer to it in terms of "place", emphasizing the "power of place" and advocating a "topological" or "topographical turn" in the context of a larger "spatial turn". In this chapter, we will first give a short introduction to *place* and *situatedness* as problems in contemporary philosophy and science (1), in order to roughly sketch the historical context in which the problem has emerged, and indicate some parallels between the different disciplinary fields. In a second step, we will turn to the concepts themselves and provide a preliminary reflection on their basic relation to each other and to other concepts (2). Finally, we will briefly present the contributions to this volume in the light of the conceptual structure unfolded in the preceding sections and evidence some maybe surprising connections between them (3).

T. Hünefeldt (✉)
ECONA – Interuniversity Center for Research on Cognitive Processing in Natural and Artificial Systems, Sapienza University of Rome, Rome, Italy

Interdisciplinary Research Group on "Philosophy of Place", Catholic University of Eichstätt-Ingolstadt, Eichstätt, Germany
e-mail: thomas.huenefeldt@uniroma1.it

A. Schlitte
Institute of Philosophy, Johannes-Gutenberg University Mainz, Mainz, Germany

Interdisciplinary Research Group on "Philosophy of Place", Catholic University of Eichstätt-Ingolstadt, Eichstätt, Germany
e-mail: annika.schlitte@uni-mainz.de

T. Hünefeldt, A. Schlitte (eds.), *Situatedness and Place*, Contributions To Phenomenology 95, https://doi.org/10.1007/978-3-319-92937-8_1

1 Place and Situatedness as Problems in Contemporary Philosophy and Science

If we start our considerations with the term "place", we can trace the recent interest in the topic back to a major problem field in twentieth century philosophy that deals with the factual conditions of human existence and its philosophical reflection.

In many respects, philosophers of the twentieth century accounted for the spatio-temporal contingency of human life. While phenomenology approached this question primarily from the perspective of perception, – emphasizing the horizontality and perspectivity as well as the embodiedness of our experience –, the dimension of history took on a more important role in the hermeneutic tradition. At the same time, philosophers of culture turned their attention towards cultural symbolic systems, which are already present for the individual, conditioning its access to the world. Finally, one can find an analogous movement in pragmatism and in the philosophy of language, which drew attention to the fact that human actions and all statements about the world are involved in a precedent language and living praxis.

Depending on the respective philosophical tradition, there are different terms used in describing the factual conditions of our being in the world, most of them related to space and time. For example, Husserl uses the term "situation" to describe a fundamental structure of our lifeworld. In unpublished manuscripts from his estate, he describes how the experience of the lifeworld takes place in concrete situations, which together form the horizon of meaning for all possible actions and fit into a comprehensive "all-situation" (Husserl 2008). While phenomenology is mainly interested in the situation of perception and knowledge, there is also a very practical meaning attached to the term "situation", which comes into view in hermeneutics and existential philosophy. One could think of Heidegger's situation of resoluteness, Sartre's situation of freedom or Jasper's "ultimate situations" ("*Grenzsituationen*"). Jaspers writes: "I am in my world only as in my situation." (Jaspers 1932, p. 56).

At first sight, "situation" seems to refer mainly to the *temporal* conditions of human existence in some of the examples (especially if speaking of a "historical situation"), but the term also implies a spatial dimension. Besides the immense discourse about history and temporality in twentieth century phenomenology and hermeneutics, there is also a lively discussion about the *spatial* conditions and, consequently, about "space" and "place", which is connected to questions of "situation", "history" and "temporality". Similar to the distinction between experienced time and objective time, we can observe a tendency to make a difference between the abstract geometric space and lived or experienced space (cf. Bollnow 2011; Ströker 1987), which led to further reflections, e.g. on the body-space in Merleau-Ponty (2012). The early Heidegger of *Being and Time* interpreted human existence mainly in temporal terms, whereas in his later thought he stressed the importance of space and, especially, place by developing the idea of a "topology of being" (cf. Malpas 2006).

In the wake of Heidegger, several philosophers, but also human geographers (cf. Tuan 2011) and social scientists (cf. Hubbard and Kitchin 2010) have carried on this thread and focused on the significance of "place" for human existence, thereby distinguishing it from "space" as an abstract scientific concept.[1] Together with the rising interest in space in the social and cultural sciences, "place" has become a productive interdisciplinary research paradigm over the last 20 years.[2] Within the context of the "spatial turn" in the cultural and social sciences, however, the differentiation between space and place is often ignored. Following discussions on postmodernity in the social sciences, place is often regarded as a "social construct" (Harvey 1996, p. 293 f.) and used as an instrument for the analysis of social and cultural phenomena, often with a political impetus (cf. Massey 1994). Against this, authors like Edward Casey and Jeff Malpas have made several attempts to do justice to place as a philosophical concept which is not only a methodological tool, but the basis of a new philosophical approach (cf. Malpas 1999; Casey 2009).

If we turn our attention to "situatedness", it is remarkable that this term has been discussed broadly in very different traditions during the last 30 years. In the framework of cultural theory, referring to one's situatedness became popular as a rhetoric at the turn of the twenty-first century, as David Simpson pointed out (cf. Simpson 2002). In pedagogy, the adjective "situated" has been used to describe an approach called "situated learning", which stressed the meaning of the social context for cognitive processes (Lave and Wenger 1991). Finally, in the cognitive sciences a highly heterogeneous research movement has developed under the heading "situated cognition", which emphasizes the "dependence" of cognition "on the situation or context in which it occurs" (cf. Robbins and Aydede 2009a), a neglected aspect within the traditional, computationalist or connectionist frameworks of cognitive science. Against the background of the historical development of the cognitive sciences, then, the "situated cognition" movement is not merely an expression of an increased attention to the dependence of cognition on the situation or context in which it occurs, but rather the symptom of a paradigm crisis in the cognitive sciences. In fact, the question as to where and how cognition is "situated" is ultimately – and quite literally – the question as to how a cognitive system is to be *defined*. Accordingly, the "situated cognition" movement may be considered as the expression of the heterogeneous search for an alternative paradigm that is capable of answering those questions.

The different approaches belonging to this movement draw on various inspirational sources which include philosophical phenomenology and philosophical pragmatism, ecological psychology and social and cultural psychology, general system theory and the theory of complex systems, cognitive robotics, and cognitive neurosciences (cf. Clancey 2009; Gallagher 2009). From all these different research areas, independently of each other, motifs have emerged which emphasize the importance of the dependence of cognition on the situation or context in which it

[1] Cf. Casey 2009; Malpas 1999; as a recent overview on the debate in hermeneutics see Janz 2017; for phenomenological approaches to place see Donahue 2017.

[2] As Tim Cresswell puts it: "Place pops up everywhere" (Cresswell 2015, p. 6).

occurs. The heterogeneity of the "situated cognition" movement manifests itself, besides their variety of sources, in the different ways in which the situation or context dependence of cognition is conceived. Both the concepts of "situation" and "context", on the one hand, and the concept of "dependence", on the other, can be understood in very different ways. On the one hand, the concept of situation (or context) can be understood in a *narrower sense*, so that it relates only to the body or even to the sensomotoric brain, but it can also be understood in a *wider sense* that includes the natural and social environment. On the other hand, the concept of dependency can be understood either as a *constitutive* dependency, so that cognition can occur only in certain situations or contexts, or in the sense of a merely *causal* dependency, so that cognition is influenced by certain situations or contexts, but can occur also independently of these situations or contexts. Accordingly, the "situated cognition" movement encompasses a broad spectrum of different strains, which are characterized by labels such as "embodied cognition", "embedded cognition", "distributed cognition", "extended mind", etc.[3]

From the brief overview of the questions this research movement addresses, it should have become plausible to draw a connection to the topic of place in continental philosophy. In fact, what all these approaches have in common – despite the unquestionable fundamental methodological differences – is the view that as human beings we are always already situated in the world and embedded in specific contexts, as phenomenology and hermeneutics have been emphasizing since the beginning of the twentieth century. Therefore, a reductionist view that sees the mind as isolated from the rest of the body and the environment is rejected both in the situated cognition movement and in the hermeneutic and phenomenological tradition.

As a consequence, the interrelatedness of world and mind comes into view. Before an action or a thought begins, we are always already in a specific place, in a cultural and historical setting, in a natural environment, in a context of meaning, in a biographical and intellectual situation. Furthermore, hermeneutic and phenomenological approaches stress the fact that philosophy itself occurs in the world, and thus it has to take its own involvement into contexts of meaning and action into account. Instead of withdrawing from the world of contingency and facticity, the place-boundedness, historicity and perspectivity can therefore be seen no longer as obstacles and restrictions, but as the enabling conditions of knowledge (cf. Malpas 2015). Similarly, within the situated cognition movement, the situation and context dependency of cognition is not regarded as a deficit, but as exactly the way how cognition works.

Despite the convergence described above, there are some profound differences between the philosophical reflection on place and the approaches labelled as situated

---

[3]An overview on the "situated cognition" movement and the strains subsumed under this major term is provided by the following monographs and volumes: Barrett 2011; Clancey 1997; Gallagher 2005; Kirshner and Whitson 1997; Menary 2010; Mesquita et al. 2011; Robbins and Aydede 2009b; Shapiro 2011.

cognition. Most obviously, the situated cognition movement advocates a naturalism which is the target of the phenomenological critique of the natural sciences, emphasizing the importance of the first-person-perspective against the predominant third-person-view. The distinction between space as an abstract scientific concept and place as belonging to the lifeworld experience can still be seen in this tradition and will therefore be skeptical towards similarities with the cognitive sciences. Moreover, although both movements share some sources (e.g., the early Heidegger, pragmatism, Merleau-Ponty), there are differences with regard to the influence of postmodern philosophy or the late Heidegger, which both contribute to the philosophical study of place,[4] but do not play any role for the situated cognition movement.

Besides the questions of influence or tradition, there are also controversial points with regard to the philosophical status of the concepts used on both sides. On the one hand, the emerging "philosophy of place" can be seen as just another way to conceptualize the spatio-temporal contingency of human life and thus as overlapping with the problem of "situatedness". On the other hand, the philosophers in question make an additional claim by assuming a certain priority of place over time and space – in experience, but even also in being.[5] In this sense Jeff Malpas has argued that place must be seen as a structure that underlies even the distinction between subject and object, when he says that

> place cannot be reduced to any one of the elements situated within its compass, but must instead be understood as a structure comprising spatiality *and* temporality, subjectivity *and* objectivity, self *and* other. Indeed, these elements are themselves established only in relation to each other, and so only within the topographical structure of place. (Malpas 1999, p. 163)

However, it remains open to debate how the notion of an ontological primacy of place is to be conceived and in how far this notion can be made fruitful beyond philosophy.

As we have seen from the last point, a basic difference between the two fields, can be found in the use of terminology. The close, but not yet totally clear connection between being situated and place becomes obvious in Casey's statement: "To be in the world, to be situated at all, is to be in place" (Casey 2009, p. XV). Hence, how can we understand the relation between "place" and "situatedness"? We will now turn to a preliminary reflection on the conceptual interrelations between "place" as the core concept of the philosophy of place, and "situatedness" as the key term in the "situated cognition" movement.

---

[4] For the influence of postmodern philosophy to the preoccupation with place see Casey 2013, pp. 285–330.

[5] Cf. Casey 2009, p. 313: "Yet the priority of place is neither logical nor metaphysical. it is descriptive and phenomenological"; whereas later he states: "The priority of places is also ontological" (ibid.). On the ontological priority of place cf. Malpas 1999, and his contribution to this volume.

2 Rethinking Situatedness and Place – Some Basic Considerations

In order to be properly grounded and intelligible, any attempt to rethink the concepts of situatedness and place should start from the linguistic context in which these two terms are themselves situated and placed. In particular, it should start from an analysis of their etymology, on the one hand, and of their contemporary common sense understanding, on the other. These analyses will evidence important basic relationships with other concepts, which will define the conceptual space of any intelligible concept of situatedness and place and which will thus allow to account for most of the critical distinctions that are referred to in contemporary thought on situatedness and place.

From an etymological perspective, the terms "situatedness" and "place" have quite different roots in Latin. The term "place" derives from Latin *platea* 'broad way, open space', which itself derives from Greek *πλατεῖα (ὁδός)* 'broad (way)', the feminine of *πλατύς* 'broad'; by contrast, the verb "situate" in the term "situatedness" derives from Late Latin *situatus*, past participle of the verb *situare* 'to place', which itself derives from Latin noun *situs* 'situation, position' (cf. English "site"), which derives from the past participle *situs* of the verb *sinere* 'to set down, leave, let'.[6] In other modern Indo-European languages, these different Latin roots have given rise to terms whose meanings more or less overlap with, but are also distinctly different from those of the English terms "situatedness" and "place".

As to "place", Italian *piazza* and Spanish *plaza* generally refer almost exclusively to an open public area in a town or city. They thus retain more of the etymological sense but do not cover the other meanings of the English term. Similarly, French *place* and German *Platz* cover the meaning of the Italian and Spanish terms, but only some of the other meanings of the English term, namely in particular those meanings which refer to something's proper or assigned position relative to others, or to some available space which may be occupied or which leaves "room" for movement. Thus, neither the Italian or Spanish terms nor the French or German terms cover that particular meaning of the English term "place" which refers to that 'wherein' one or more things and events 'take place'. In other modern Indo-European languages, in fact, this particular meaning, which is particularly relevant both in the context of contemporary thought on "place", in general, and in the context of a reflection on the relationship between "situatedness" and "place", in particular, is expressed by terms with other roots, for example by Italian *luogo*, Spanish *lugar*, and French *lieu*, which all derive from Latin *locus*, and by German *Ort*, which derives from the same Germanic root as Old-English *ord* meaning basically "point" (e.g., of a sword) and therefore also "line of battle, forefront" or "beginning, origin, source".[7] Especially the German term "Ort" thus involves

[6]Cf. the entries "place" and "situate" in Onions et al.'s (1983) *Oxford Dictionary of English Etymology* and in Klein's (1971) *Comprehensive Etymological Dictionary of the English Language*.

[7]Cf. the entry "Ort" in *Digitales Wörterbuch der deutschen Sprache*: https://www.dwds.de/wb/Ort#1, and the entry "ord" in Bosworth's (1898) *Anglo-Saxon Dictionary*.

connotations which run rather counter to those of the English term "place", in particular in so far as it does not require the preposition "in" (*in*) but "at" (*an*).

As to "situatedness", this term is not easy to render in other Indo-European languages and generally needs to be paraphrased by recurring either to the past participle of verbs derived from the Latin verb *situare*, as in Italian *situato*, Spanish *situado*, French *situé*, or German *situiert*, which all have connotations that are more or less slightly different from those of English *situated*, or to terms derived from other roots. The term "situatedness" is thereby to be distinguished from the term "situativity", which is a nominalization of an adjective derived from the same Latin root and which has corresponding terms in other modern Indo-European languages terms (cf., e.g., Italian *situatività*, Spanish *situatividad*, French *situativité*, or German *Situativität*). In fact, although it is sometimes used synonymously with "situatedness", the term "situativity" emphasizes rather the affectability by, or causal dependence on, changing situations than the rootedness in, or constitutive dependence on, a particular place, whereas the term "situatedness" emphasizes rather the latter than the former. The term "situatedness" thus conserves more strongly than the term "situativity" the original meaning of their common root in the Latin noun *situs*, which refers to the place where something is settled.

Considering the etymology and semantic field of the terms "situatedness" and "place", the contemporary common sense understanding of these terms may be characterized roughly in the following way: "situatedness" refers to the fact that, or to the way in which, something is in some place, while "place" refers to that wherein something is or may be situated. These rough and necessarily approximate characterizations suggest several important points concerning situatedness, place, and the relationship between them:

First, neither situatedness, nor place, nor the relationship between them can be conceived without reference to something which may be supposed to be situated or to be in some place. In fact, as situatedness cannot be conceived without reference to something which may be supposed to be in some place, so place cannot be conceived without an at least implicit reference to something which may be supposed to be situated somewhere, and *a fortiori* the relationship between situatedness and place cannot be conceived without reference to something which may be supposed to be situated in some place.

Second, the relationship between situatedness and place is an intrinsic relationship which is characterized by a reciprocal implication: situatedness implies something's being in some place and place implies the possibility of something's being somehow situated therein. Of course, there are different places in which something may be situated and different ways in which it may be situated therein, but the plurality of places and of ways of being situated therein does not affect the intrinsic link between situatedness and place.

Third, this intrinsic relationship between situatedness and place implies a priority of place with respect to situatedness which is analogous to the "priority" (πρῶτον εἶναι) that Aristotle granted to place with respect to "the things in it" (τὰ ἐν αὐτῷ): as for Aristotle place can exist even without the "things in it", whereas these "things" cannot exist without place,[8] so there can be place without actual situatedness, whereas there cannot be situatedness without place.

Already for Aristotle, the latter point is eventually based on the common-sense supposition that "everything is somewhere and in place" (πάντα εἶναί που καὶ ἐν τόπῳ), or more precisely that "every sensible body is in place" (πᾶν σῶμα αἰσθητὸν ἐν τόπῳ).[9] Now, this reference to a totality of "things", regardless of how they are specified, implies an important conceptual differentiation concerning both "situatedness" and "place". In fact, "place" may refer not only to that wherein something is or may be situated, i.e. to some particular place in the world, but also to that wherein everything is and must be situated, i.e. to the all-encompassing place of the world. Analogously, "situatedness" may refer not only to the contingent relationship to some particular place in the world, but also to the necessary relationship to the all-encompassing place of the world. In other words, "everything" is *necessarily* situated not only in its own place, i.e. in the particular place defined by that what takes place in it, but more generally also in the all-encompassing place of the world wherein it takes place, while it is only *contingently* situated in a particular place relative to the particular places of other "things", i.e. to the particular places wherein these other "things" take place. Of course, the *necessity* and *contingency* that are here referred to are not physical but metaphysical, since the "things" that are here referred to are not further specified. Accordingly, the aforementioned relationships may be supposed to hold for any conceivable world.

Starting from the common-sense understanding of situatedness and place, it thus results that any intelligible concept of situatedness and place implies not only the distinction between place defined in terms of that what is or may be situated in it and place defined in terms of its relation to other places, but also the distinction between a particular place in the world and the all-encompassing place of the world. The concepts of situatedness and place are thus necessarily related to the concept of world.

Now, common sense distinguishes between different kinds of worlds, i.e. in ordinary language the common-sense concept of world is specified in different ways. For example, it is common to distinguish:

- ***the*** **"real"** world vs. ***a*** **"represented"** world, i.e. the world that is actually given and lived vs. any world that is in some respect "different" from the world that is actually given and lived (e.g., a past or future world);
- ***the*** **"own"** world vs. ***an*** **"other's"** world, i.e. the world that is actually given and lived insofar as it is represented to be the world of a particular subject among others (e.g., my world), vs. the world that is actually given and lived as it is represented to be for another subject (e.g., your world);

[8] Cf. Aristotle (1957), *Physics*, Book 4, 208b-209a. Aristotle attributed this supposition – which basically amounts to his thesis that place is a category – to "most people", i.e. literally to "the many" (οἱ πολλοί, cf. 208b), or even to "all" (πάντες, cf. 208a).

[9] Ibid.

- ***a* "subjective"** world vs. ***the* "objective"** world, i.e. the world that is actually given and lived as it is represented to be for a particular subject, (e.g., my or your world), vs. the world that is actually given and lived as it is represented to be independently of any particular subject (e.g., the world as it is represented by science or by the naïve realist);

Together with the distinction between a particular place in a world and the all-encompassing place of the world, and with the distinction between a particular place defined in terms of that what is or may be situated in it and a particular place defined in terms of its relation to other places, these basic distinctions concerning the all-encompassing place of the world define the conceptual space of any intelligible concept of situatedness and place. Together, they should therefore allow to account for most of the critical distinctions referred to in contemporary thought on situatedness and place, for example the distinctions between "place" and "space", on the one hand, and between "place" and "position" or "site", on the other.[10]

In fact, many of these critical distinctions may be described in terms of these common-sense distinctions. For example, the distinctions between "place" and "space", on the one hand, and between "place" and "position" or "site", on the other, may be described in terms of (1) the distinction between the all-encompassing place of the world that is actually given and lived, on the one hand, and the all-encompassing place of a supposed, i.e. abstractly represented objective world, on the other; (2) the distinction between a particular place in the former world and a particular place in the latter world; and (3) the distinction between a particular place insofar as it defined by that what takes place therein and a particular place in so far as it is defined by its relation to other places.

As the concepts of situatedness and place are necessarily related to the concept of a world wherein takes place whatever takes place, any intelligible concept of situatedness and place must eventually be grounded in the world that is actually given and lived, i.e. in that what in the phenomenological tradition is called the "life-world". In other words, situatedness in a particular place within the all-encompassing place of the world that is actually given and lived is the paradigm for any consistent and coherent understanding of situatedness and place. Now, among the "things" that are situated in a particular place within that world there is also a "thing" that may be, and by normal human beings generally is, identified as oneself, i.e. as one subject, or conscious being, among others. The paradigm of the world that is actually given and lived does therefore not only imply a distinction between different ways of situatedness in different kinds of particular places within the all-encompassing place of that world, but it also contains the ground for distinguishing between different subjective worlds and the one objective world, in which oneself and other subjects are supposed to coexist as particular "things" among others.

Not only the distinction between different subjective worlds and the one objective world, but virtually any distinction between different kinds of worlds implies reference to mind, insofar as mind must be supposed as that what constitutes

[10] Cf., for example, Casey (2009).

different subjective worlds and what allows to represent worlds different from the world that is actually given and lived. Accordingly, all those distinctions concerning the concepts of situatedness and place which are eventually based on the distinction between different kinds of worlds rely at least implicitly on reference to some concept of mind. In particular, this is true for many of the aforementioned critical distinctions referred to in contemporary thought on situatedness and place, for example for those relying on the distinction between the all-encompassing place of the world that is actually given and lived, on the one hand, and the all-encompassing place of a supposed, i.e. abstractly represented objective world, on the other.

These relationships result in a close entanglement between situatedness and place, on the one hand, and mind, on the other. On the one hand, mind cannot be conceived without reference to situatedness and place, insofar as it is supposed to take place at all, and it cannot be conceived without reference to its situatedness in particular places, insofar as a plurality of minds is supposed. On the other hand, both situatedness and place may be conceived without reference to mind, insofar as they may be referred to the world that is actually given and lived, regardless of how that world is represented, and thus to any conceivable world; but they cannot be conceived without reference to mind, insofar as they are *differently* referred to *different* worlds, for example if one of these terms (e.g., "place") is used preferentially with reference to the world that is actually given and lived, rather than to the abstractly represented objective world supposed by science. Due to its implicit reference to mind, such differential use of these terms involves the risk of an either subjectivistic or objectivistic misunderstanding of the positions thereby intended.

To sum up, the preceding considerations have provided a series of elements that may help to analyze and understand the relationship between the contemporary thought on "place", on the one hand, and on "situatedness", on the other. Starting from an analysis of the etymology and of the common-sense understanding of "situatedness" and "place", we have first evidenced the intrinsic relationship between these two concepts and thus between the conceptualizations associated with them. We have then evidenced their necessary relationship with other common-sense concepts such as "thing" and "world", which are involved in some basic distinctions that define the conceptual space of any intelligible concept of situatedness and place. In particular, we have distinguished between:

- place defined in terms of that what is situated in it, on the one hand, and place defined in terms of its relation to other places, on the other;
- particular places in the world, on the one hand, and the all-encompassing place of the world, on the other;
- different kinds of worlds, i.e. different ways of representing the all-encompassing place of the world.

Given these distinctions, we have then argued that any intelligible concept of situatedness and place must eventually be grounded in a phenomenological analysis of the world that is actually given and lived, and in a phenomenological account of the distinction between different kinds of worlds. On this basis, we have finally hinted at the close entanglement between situatedness and place, on the one hand, and mind, on the other.

3 The Contributions to This Volume

Since this volume focusses on the relationship between the concept of "place" in contemporary phenomenological and hermeneutic philosophy and the concept of "situatedness" in contemporary cognitive science, many other ways of understanding "place" and "situatedness", especially regarding cultural theory and gender theory, are not represented in this volume. In particular, this focus implies a shift from the use of the term "situatedness" in cultural theory to its use in cognitive science, where the relationship with philosophical concerns about "place" and "situation" is not evident at first sight. A main objective of this volume is in fact to open a dialogue between two radically different but themselves heterogeneous research traditions and to facilitate references across the conventional disciplinary divisions. In the last section of this introductory chapter, we will therefore briefly present the contributions to this volume in the light of the conceptual structure unfolded in the preceding sections and evidence some maybe surprising connections between them.

The first three contributions, by Edward Casey, Jeff Malpas, and David Seamon, approach the topic from the perspective of a philosophy of place. Their papers share the attempt to provide a systematic reflection on the terms being used in both traditions – most prominently place, situation, and situatedness – and combine those considerations with methodological thoughts, especially on the role of phenomenology for the analysis of place, which is very revealing for the internal diversity and complexity of the philosophical discourse on place and situation. How can we approach place and situation from a philosophical point of view – what role do phenomenology and hermeneutics take in this enterprise? The opening papers of this volume thus lay the ground for the dialogue with the research traditions represented by the following, somewhat more specific chapters.

In his paper "Place and Situation", Edward Casey suggests an account of place and situation which both points out their similarities and their differences. He starts from the common sense understanding of those terms, seeing place as location ("*where something is to be found*"), and situation as the way "how things are *disposed in place*". He then employs examples from everyday life, which illustrate the complex intermingling of the two and call for a philosophical differentiation. It thus becomes obvious in his analysis, that place precedes situation, insofar as it serves as a minimal condition for situation. The core of the concept of place is its interrelatedness with the body ("*no place without body; and no body without place*"), which functions as both occupant and animator of place, as Casey points out. Besides that, the dimensions of history and event are only of minor importance for place, although they always appear in specific places. They are brought to place by situation: In contrast to the mono-elemental, essentially singular and unique character of place, situation is described as essentially plural and multi-elemental, and decisively characterized by its event-like-character. Since it is no such unique whole as place, situation is more open to human action, as it seems. Situation, then, is realized in acts of synthesis, imagination, and freedom, and thus it exceeds place, although place is still needed as a "material condition of possibility".

Jeff Malpas' paper "Place and Placedness", by contrast, focusses not on the distinction between place and situation, but on the difference between place (and situation) on the one hand, and placedness (and situatedness) on the other. Whereas placedness is a characteristic of what is being placed, place is that to which what is being placed is related. According to Malpas, both concepts are too often not distinguished appropriately, especially when place is understood as dependent on human apprehension. Such a view reduces place to placedness, which would imply that place somehow belongs to that what is placed in it (i.e., the subject), whereas in his view place precedes any form of being placed. Malpas sees this false "subjectivism" at work in both philosophical reflections on place and in cognitive science stressing the situated and embodied character of cognition. Especially the idea of an "egocentric space", explaining orientation independently from any "objective" space, becomes a target of criticism, as well as the wide-spread understanding of place as "meaningful space". However, place shall not be understood as objective either, but Malpas claims that the subjective, the intersubjective and the objective are part of one overall structure, which he calls "triangular" using Davidson's term. Accordingly, place is not reducible to either the first- or the third-person-perspective. This is an interesting point, because the relation between phenomenology and cognitive science is often reconstructed in terms of this difference of perspectives (cf. chapter "The Place of Mind").

Malpas' verdict on the preference for placedness/situatedness over place concerns the early Heidegger as well as Merleau-Ponty, and with them many authors referring to them across the disciplines. Against a phenomenological approach to place, Malpas stresses the importance of the hermeneutical view and makes the case for an ontological understanding of place. Taking his criticism seriously, the relation between body and place cannot be as fundamental as Casey or Seamon see it (cf. chapters "Place and Situation" and "Merleau-Ponty, Lived Body, and Place: Toward a Phenomenology of Human Situatedness"); instead of being correlative with the body, place becomes correlative with being. In a way, the ontological difference between being and beings is repeated in the distinction between place and places; place as such and the concrete places always occupied by that which is being situated in them.

Differently from Malpas (chapter "Place and Placedness") and more explicitly than Casey (chapter "Place and Situation"), David Seamon puts much emphasis on the methodological question of how to deal with place within a phenomenological framework. For him, the task of phenomenology is to make the hidden dimensions of our everyday place-experience explicit. By referring to Merleau-Ponty and Casey, Seamon identifies the body-subject as the key-figure in our experience of the world, which is characterized by bodily and time-space routines. Combining the interpretation of Merleau-Ponty with an example taken from literature and a first-person-description of everyday phenomena, Seamon suggests a method which uses the contrast between texts on different descriptive levels. Seamon calls this method "triangulation", although in a sense different from the one employed by Malpas.

In his chapter "Situating Interaction in Peripersonal and Extrapersonal Space: Empirical and Theoretical Perspectives", Shaun Gallagher focuses on "the relationship between embodied intersubjective interactions and the way that such interactions shape and are shaped by the way we experience space", both "the reachable (peripersonal) space around us" and "the more distant (extrapersonal) space beyond our immediate reach". Gallagher's chapter thus represents a current within the "situated cognition" movement that is known as "enactivism". Roughly, enactivist theories maintain that cognition is shaped by the dynamic interaction between an acting organism and its environment. In line with such theories, Gallagher holds that "the way we experience space" is shaped by our embodied interaction with both our physical and our social environment. Such an enactivist view is strikingly similar to, but also characteristically different from, the view expressed in Casey's formula "no place without body; and no body without place" (see chapter "Place and Situation"), which refers – as Seamon accurately summarizes it – to the "co-constitution of lived bodies and place" (see chapter "Merleau-Ponty, Lived Body, and Place: Toward a Phenomenology of Human Situatedness"). In fact, as for Gallagher "the way we experience space" is shaped by our embodied interaction with our physical and social environment, so for Casey "[p]lace [...] is the outcome of bodily engagement" (see chapter "Place and Situation"), and as for Casey bodily engagement supposes, and is conditioned by, the place wherein it takes place, so Gallagher wouldn't deny that our embodied interaction with our physical and social environment supposes, and is conditioned by, the place wherein it takes place. The two views seem to differ mainly in how they conceive the relationship between the first- and the third-person perspective on spatial relationships, and thus between phenomenology and empirical science. Whereas enactivists such Gallagher try to account for the phenomenological, first-person perspective on spatial relationships ("the way we experience space") in terms of empirical, third-person perspective on our embodied interaction with our physical and social environment ("embodied intersubjective interactions"), philosophers of place such as Casey and Seamon focus on the phenomenological, first-person perspective on spatial relationships and dismiss the empirical, third-person perspective on spatial relationship as derivative, abstract, and thus alienating.

People's embodied interaction with their physical and social environment is also, though with a different focus, a central theme in William J. Clancey's chapter "Spatial Conception of Activities: Settings, Identity, and Felt Experience". In fact, Clancey approaches this topic from the perspective of the "socio-technical approach to developing technology", in which "social scientists and computer scientists ground designs of automation (e.g., software, devices, vehicles) in ethnographic studies of how people interact with each other, tools and representations (e.g., computer displays), and their environment (e.g., facilities)". The "socio-technical approach to developing technology" represents an important part of the "situated cognition" movement in so far as "[t]he 'situated' perspective in the analysis and design of socio-technical systems reveals how people conceive of activities as social-interactive settings". Accordingly, a key purpose of Clancey's chapter is to elucidate "the multi-dimensional [...] nature of settings with examples from ethnographic studies of robotically mediated field science on Mars and analog expeditions on Earth". Due its "multi-dimensional nature", which includes not only "both personal and social, both physical and conceptual," but also both "emotional" and "interactive", and both "aesthetic" and "practical" aspects, Clancey's notion of "setting" has much in common with the cultural and social notion of "place" that has been emphasized in the wake of the "topological turn". Notably, Clancey repeatedly suggests that these two notions "blend" or explains the one in terms of the other. Furthermore, the "ethnographic" approach employed in Clancey's "socio-technical approach to developing technology" eventually relies on multi-dimensional phenomenological analyses such as those described in Seamon's chapter, in so far as they provide data for making general statements about people — in particular their socio-cognitive experience, practices, and values. However, for the reasons mentioned above, the step towards an empirical third-person perspective on spatial relationships that is implied in making such general states about people is precisely what Seamon and other phenomenologically inspired philosophers of places find problematic.

Whereas the first five contributions to this volume are by authors representing either the "situated cognition" movement or the "philosophy of place", the following five contributions are by authors who are not clearly associated with either one or the other research tradition, but work on problems inspired by those traditions.

In his paper, "The Place of Mind", Thomas Hünefeldt offers a framework for a systematization of different ontological answers to the question as to where mind takes place. By addressing this question, he opens up a perspective that is able to include recent debates in cognitive science as well as traditional philosophical oppositions. After going through the different logical possibilities, where mind can be located (nowhere, everywhere, somewhere), he reassesses those different answers from the perspective of phenomenological experience, which he regards as "epistemologically foundational". The phenomenological concept of mind serves as a touchstone for the theories described above, and provides an explanation for the surprising variety of answers, since it "implies an intrinsically ambiguous relationship between mind and world". From the phenomenological perspective, mind is on the one hand a place in which everything takes place, since it constitutes a subjective world, but on the other hand it is also something that takes place in particular places/entities in the world. Although having started on a more abstract level of a distinction between different types of ontology, Hünefeldt's paper ends at a point that refers back to Casey's and Seamon's argument for a phenomenological account of place, when he, like them, insists on the foundational character of phenomenology.

In her chapter "Place and Positionality – Anthropo(topo)logical Thinking with Helmuth Plessner", Annika Schlitte explores the role of place in Helmuth Plessner's philosophical anthropology. In particular, she argues that Plessner's approach may be understood as a "twofold 'implacement' of man", which results in the paradoxical concept of man's "eccentric positionality". Plessner's approach is particularly interesting in the context of a reflection on the relationship between the philosophical discussion on place and the situated cognition movement in cognitive science because it represents a peculiar way of mediating between the third- and the first-person perspective on man's place in the world. In fact, Plessner mediates between these perspectives by distinguishing different "levels of the organic", leading from inanimate objects to plants, animals, and humans. For Plessner, each of these four levels is characterized by a particular "relation to space", which only in the case of animals and humans includes a first-person perspective on space. By thus allowing for non-human "subjective worlds" (cf. chapter "The Place of Mind"), Plessner's account points at a conception of place that goes beyond the "anthropocentrism" characteristic of many contemporary discussions on place.

Tsutomo Ben Yagi's paper, "Birth in Language: The Coming-to-Language as a Mark of Non-difference in Gadamer's Hermeneutics", deals with our situatedness in language. Rejecting Waldenfels' critique of Gadamer, which states that hermeneutics could not account for radical foreignness, Yagi follows Gadamer in underlining the importance of the mother tongue as something we cannot get beyond, as something that lies beyond the logic of own and foreign and thereby any "logic of difference". The fact that Gadamer speaks of language as "home" is revealing here. Both, the place we find ourselves always already in, and the mother tongue we are always already mastering when we start reflecting on it, belong to the seemingly familiar but puzzling conditions of our being in the world, which escape from view too easily.

Tobias Holischka's chapter "Virtual Places as Real Places: A Distinction of Virtual Places from Possible and Fictional Worlds" focusses on an increasingly important and so far largely neglected aspect of the discussion on place: virtual places and the situatedness therein. Holischka first investigates the ontological status of virtual places by connecting them to our lifeworld and by distinguishing them from fictional places and merely (logically) possible worlds. In order to adequately describe the variety of phenomena that may be considered as virtual places, he then distinguishes between three "instances of virtual placement", which he calls "trans-placement", "in-placement", and "re-placement". Finally, he illustrates what it is like to be living in a virtual world by the example of the video game *Minecraft*. Holischka's discussion of virtual places obviously has a strong relation to Clancey's discussion of the scientists who are exploring Mars (cf. chapter "Spatial Conception of Activities: Settings, Identity, and Felt Experience"), for Mars is a real place that (at least so far) can be experienced only through a kind of virtual environment provided by computer media (including photographs & robotic systems). Furthermore, as illustrated in Holischka's discussion of *Minecraft*, virtual environments enable interactivity and thus constitute a particular kind of "setting" in Clancey's sense of that term.

Finally, in the chapter "Situated Anxiety: A Phenomenology of Agoraphobia", Dylan Trigg argues for the situated quality of anxiety, in general, by focusing on a particular kind of anxiety, namely agoraphobia, which is commonly thought of as a spatial anxiety, especially a fear concerning public places. Following a critical analysis of Heidegger's "placeless, disembodied, and individualist" account of anxiety, on the one hand, and a critical discussion of the prevalently "biological" or "cognitive" understanding and treatment of anxiety in contemporary clinical psychology, on the other, Trigg sketches the move "towards a situated account of agoraphobia" which emphasizes the "bodily, spatial, and intersubjective dimensions" of this kind of anxiety. Though this move draws not only on autobiographical reports by agoraphobic individuals, but also on clinical reports by psychologists and physicians, it eventually amounts to a "phenomenological analysis of agoraphobia". Despite Trigg's explicit critique of "third-person", and in particular "cognitive", accounts of anxiety, his move "towards a situated account of agoraphobia" is not at all incompatible but fits quite well with "enactivist" approaches such as that proposed by Gallagher (cf. chapter "Situating Interaction in Peripersonal and Extrapersonal Space: Empirical and Theoretical Perspectives"). In fact, rather than being a critique of "cognitive" approaches, in general, it is a critique of traditional cognitive approaches that do not emphasize the enactive "situatedness" of cognition.

The connections that we have evidenced between the different chapters of this volume do of course represent only a somewhat arbitrary part of the connections that might possibly be identified, and they could only be hinted at very briefly. Nevertheless, they may be a starting-point for further and more detailed explorations.

References

Aristotle. 1957. *Physics Book 1–4*. Trans. P.H. Wicksteed and F.M. Comford. Cambridge: Harvard University Press.

Barrett, L. (2011). *Beyond the brain: How body and environment shape animal and human minds*. Princeton: Princeton University Press.

Bollnow, O.F. 2011. *Human Space*. Trans. Christine Shuttleworth and ed. Joseph Kohlmaier. London: Hyphen Press.

Bosworth, J. 1898. *An Anglo-Saxon Dictionary: Based on the Manuscript Collections of the Late Joseph Bosworth*, ed. T.N. Toller. Oxford: Clarendon Press.

Casey, E. 2009. *Getting Back into Place. Toward a Renewed Understanding of the Place World*. 2nd ed. Bloomington: Indiana University Press.

———. 2013. *The Fate of Place. A Philosophical History*. 2nd ed. Berkeley/Los Angeles/London: University of California Press.

Clancey, W.J. 1997. *Situated Cognition: On Human Knowledge and Computer Representation*. New York: Cambridge University Press.

———. 2009. Scientific antecedents of situated cognition. In *The Cambridge Handbook of Situated Cognition*, ed. P. Robbins and M. Aydede, 11–34. New York: Cambridge University Press.

Cresswell, T. 2015. *Place. An Introduction*. 2nd ed. Malden/Oxford: Blackwell Publishing.

Donahue, J., ed. 2017. *Place and Phenomenology*. London/New York: Rowman and Littlefield.

Gallagher, S. 2005. *How the Body Shapes the Mind*. Oxford: Oxford University Press.

———. 2009. Philosophical antecedents of situated cognition. In *The Cambridge Handbook of Situated Cognition*, ed. P. Robbins and M. Aydede, 35–52. New York: Cambridge University Press.

Harvey, D. 1996. *Justice, Nature and the Geography of Difference*. Cambridge, MA/Oxford: Blackwell.

Hubbard, P., and R. Kitchin, eds. 2010. *Key Thinkers on Space and Place*. 2nd ed. London: Sage Press.
Husserl, E. 2008. *Die Lebenswelt. Auslegungen der vorgegebenen Welt und ihrer Konstitution,* Texte aus dem Nachlass 1916–1937. Hua XXXIX. Nr. 20 and 21. ed. Rochus Sowa. Dordrecht: Kluwer.
Janz, B., ed. 2017. *Place, Space and Hermeneutics* (Contributions to Hermeneutics, Vol. 5). New York: Springer.
Jaspers, K. 1932. *Philosophie, Vol. 1: Philosophische Weltorientierung*. Berlin: Springer.
Kirshner, D.I., and J.A. Whitson, eds. 1997. *Situated Cognition: Social, Semiotic, and Psychological Perspectives*. Mahwah: Lawrence Erlbaum Associates Publishers.
Klein, E. 1971. *A Comprehensive Etymological Dictionary of the English Language*. Amsterdam/London/New York: Elsevier Publishing Company.
Lave, J., and E. Wenger. 1991. *Situated Learning: Legitimate Peripheral Participation*. New York: Cambridge University Press.
Malpas, J. 1999. *Place and Experience*. Cambridge: Cambridge University Press.
———. 2006. *Heidegger's Topology*. Cambridge, MA: MIT Press.
———. 2015. Place and situation. In *The Routledge Companion to Hermeneutics*, ed. J. Malpas and H.-H. Gander, 354–366. Oxon/New York: Routledge.
Massey, D. 1994. *Space, Place, and Gender*. Cambridge, MA/Oxford: Blackwell.
Menary, R., ed. 2010. *The Extended Mind*. Cambridge: MIT Press.
Merleau-Ponty, M. 2012. *Phenomenology of Perception*. Trans. D.A. Landes. New York: Routledge.
Mesquita, B., L. Feldman Barrett, and E.R. Smith, eds. 2011. *The Mind in Context*. New York: Guilford Press.
Onions, C.T., G.W.S. Friedrichsen, and R.W. Burchfield. 1983. *The Oxford Dictionary of English Etymology*. Oxford: Clarendon Press.
Robbins, P., and M. Aydede. 2009a. A short primer on situated cognition. In *The Cambridge Handbook of Situated Cognition*, ed. P. Robbins and M. Aydede, 3–10. New York: Cambridge University Press.
———, eds. 2009b. *The Cambridge Handbook of Situated Cognition*. New York: Cambridge University Press.
Shapiro, L. 2011. *Embodied Cognition*. New York: Routledge/Taylor & Francis Group.
Simpson, D. 2002. *Situatedness, or, Why We Keep Saying Where We're Coming From*. Durham/London: Duke University Press.
Ströker, E. 1987. *Investigations in Philosophy of Space*. Trans: A. Mickunas. Athens: Ohio University Press.
Tuan, Y. 2011. *Space and Place. The Perspective of Experience*. 2nd ed. Minneapolis/London: University of Minnesota Press.

Place and Situation

Edward S. Casey

Abstract "Place" and "situation" are often confounded in everyday discourse; yet they have crucially different dimensions. Place is locatory and singular, and is the outcome of bodily engagement: to be a lived body is to be in place; and to be in place is to be there by way of body. Situation contributes scope and setting to place itself. In particular, it brings temporality and historicity to bear on place, broadening it and making it more reflective of vicissitudes to which it is subject. Situations occur primarily as events that unfold in time as well as space. They call upon acts of synthesis, imagination, and freedom in their full realization. Place and situation belong together even as they are distinguishable in these various ways.

Keywords Place · Space · Situation · Body · Event · Aristotle · Sartre

1 Place vs. Situation

At first sight, it would seem easy to distinguish place from situation: "place" would be location, *where something is to be found*; "situation" is how things are *disposed in place*. After all, there is a long tradition of considering place as a sheer locus, indeed as a *container*, as on Aristotle's view in *Physics* Book IV: "place (*topos*) is the first unchangeable limit of that which surrounds." (Aristotle 1983, p. 28 [Book IV, 212a, 20–21]) Place as a surrounding presence makes a certain amount of sense – common sense, which seeks security and stability. Things and events need a sheltering presence, a space in which to be or to become what they are: a *somewhere* in which to gain a certain stability: --hence the redundant but telling Roman expression *stabilitas loci*. As for situation, it takes care of all the rest: above all, how what is in place is arranged, especially in terms of the nexus formed by its internal and

E. S. Casey (✉)
Department of Philosophy, SUNY Stony Brook, Stony Brook, NY, USA

T. Hünefeldt, A. Schlitte (eds.), *Situatedness and Place*, Contributions To Phenomenology 95, https://doi.org/10.1007/978-3-319-92937-8_2

external relations: how its contents are related to each other, and how these contents are related to things outside its immediate location. If the paradigm of place is indeed that of a container, that of situation is that of the entire *circumstance* of that same container. This much seems intuitively evident.

Examples abound in everyday life. They range from the straightforward to the complex. On the simple side: when I'm writing this essay at a Starbucks café in Santa Barbara, the café itself is the place I'm in, and the arrangement of the tables and chairs gives the situation of which I'm a part. But more complicated cases come quickly to mind. Take that of a migrant confronted with the U.S.-Mexico border wall. What is his place? What is his situation? His place is already complicated. To begin with, he is between two places: he stands at the wall on the Mexican side but is actively imagining being on the other side; physically, he can be on only one side at a time; psychically, he is on both. In a case like this, place shows itself to be strikingly *bivalent* – quite different from Aristotle's monovalent model that accounts for only one place at a time, and then solely a physical place at that. And the situation of the migrant? It, too, is more complicated than many ordinary situations. A migrant is caught up in a complex circumstance that includes local and national and international economies; family relations (including the responsibility of the migrant to support those family members who remain in the home country); political realities (e.g., the difference that Donald Trump's attitude toward the border wall is now making in contrast with that of Barack Obama). I here point only to the most obvious parameters of a circumstance that is highly ramified and without effective limit. Immediate place qua location is only part of the migrant's situation though it is altogether essential. Being here, in this physical locale, is as Husserl insisted, "absolute": one is altogether located there, without remainder; at the same time, it is sine qua non necessary if one is to have a determinate locus. At the same time, the situation that presumes it surpasses it on all sides.

The analysis thus far might suggest that the difference between place and situation is merely, or mostly, quantitative: in situations there are more items (things and relations between things) than in a single place. But this difference is merely relative, and would allow for situation to be little more than an expanded place. On the face of it, there seems to be no definitive, much less essential, difference between place and situation. But thinking this way is to miss crucial dimensions of each of these paired terms. Let me now spell out some of these dimensions by way of brief description, taking up place first and then situation.

As for *place*: place is not a simplex; it is highly structured in its own unique way. Not only does it possess an absolute hereness that belongs to it alone; but the bearer of such non-relative locatory force is the *lived body*. This body is at once the occupant and the animator of place. In a formula I like to employ in this context: *no place without body; and no body without place.*[1] It is by the body that I am where I am – that I am in the place that I designate simply as "my place." If a place is to be considered mine, this is only because – and to the extent that – I am able to move

[1] "Just as there is no place without body… so there is no body without place" (Casey 2009, p. 104).

there to begin with and to settle into it. The settling itself can be as momentary as occupying a place in a line waiting for a bus, or as long-lasting as what I come to designate as my "home-place." Either way, the body is at once the subject of place and its animator; it is the agent who is not just aware of the place it takes up but also able to alter it in ways that suit its staying there longer.

Unless, of course, it is forced to leave that place – as with the migrant at the wall. This sobering spectacle, all too common in this era of nationalistic wall-building, reminds us that no place is to be taken for granted. No matter how long my body has been settled into a given place or how successfully, the place itself can fail to offer a viable life, as happens in the extremities of climate change where bodies are forced to move out of even the most accustomed places they are used to inhabiting: as in regions of the earth where increasing aridity is driving whole peoples out of their homes and towns in a massive movement of forced migration.

With the invocation of forced migration – whether due to economic, political, or climate reasons – we reach a point where situation impinges upon place: where we cannot hold place and situation apart as if they were pristine epicenters of human (or for that matter) animal life. Even short of a full consideration of situation by itself, situation intrudes into the fate of place in ways that cannot be ignored and that make up, along with the lived body, the complexity of place itself. Situation introduces parameters that exceed place itself, importing parameters that come from, and ramify into, a larger world than place itself can encompass.

A further factor complicatinges any putative simplicity of place is *historicity*: the fact that human beings are self-interpreting creatures who by the compilation of archives (physical or electronic) and by the use of language and other symbolic modes of expression accomplish ways of understanding themselves and the histories to which they belong in ways that are uniquely configurative and reflective both of individuals and entire groups. Thus, we talk about the places where we have been, and we write about our experiences there in diaries or letters: all of this part of the vast and subtle process of remembering place. What is otherwise bare place in physical fact is quite literally over-written with words and spoken-through in oral discourse. It may also be imaged in photography and painting. In these culturally specific, and historically rooted, ways place becomes complicated. Or rather we should say: it is always already complicated: from the start, and not only subsequently in the many aftermaths to which it gives rise.

In all these various ways, place cannot be constricted to sheer location. It is *striated* as it were, possessing many strata of differential structures and diverse significations. The strata are not superficial or merely additive; they are intrinsic to place, immanent to its very being. From the very beginning, place is complicated – folded through and through, over and under with layers that are not always easy to distinguish but that contribute to its placial identity. They contribute to our sense that a given place is indeed *that* very place: that one place, that same place. This is a sameness that is constituted by difference itself.

2 Im-plications and Com-plications

If place is com-plicated and ex-plicated by closely accruing stratifications, *situation* is im-plicated by place in turn: it folds *into* place even as it exceeds it. Situation needs a minimal sense of place to be capable of holding what is in its embrace. To be situated means, minimally, to have a place from which things and events can be coherently related to each other. Place is thus a minimal material condition for situating to occur. A placeless situation is a non-sequitur – a non-starter. It is so many angels dancing on the head of a pin. (In fact, the celebrated medieval conundrum concerned the essentiality of sheer place to existence: a debate taken up by Kant in his Inaugural Dissertation).[2] Some kind of place, some kind of locatory stretch of space, is required if situating is to occur. But a given place, no matter how complex it may become or how far it builds out from itself, is never itself a situation.

Moreover, even if a given place has several layers or dimensions, it is finally one integral whole. This is to say that given place is essentially singular. It is *mono-elemental*; its placiality as such is unique. This is why one place can never take the place of another place: "there's no [other] place like home." A situation, in contrast, is essentially *multi-elemental*: it is composed of several factors or elements, never just one. As a consequence, its uniqueness is not of central concern; being essentially plural, it mixes and merges with other situations of a like kind. Thus we say that "the living situation today is much like it was ten years ago"; but the particular places that subtend these situations are incomparable: their locations may be the same, but the exact mode of their occupation by lived bodies will differ.

Take the situation of attending a piano recital. This happens in one place (the recital hall); however structurally complex the architecture and layout of this hall may be, and no matter how many layers of meaning it bears, the recital happens *just there*, in that particular place. If I am to hear the recital in first person, I must take myself *to that place*. To be in another place at the time of the recital is to miss the recital; it is to be elsewhere. But once in the right recital hall and at the right time, I find myself in a situation with several distinguishable components: the hall, the piano, the stage, and the performer; myself, other selves, seats in the hall; ushers and other personnel. All this, at the very least, makes up the situation. If the place – the recital hall – is steadily the same, the situation in which the recital takes place is multiplex in the very midst of that sameness. The *one* hall is where a comparatively complex situation unfolds. Its sheer sameness undergirds and makes possible the multiplicity of the situation that exfoliates there.

A bare place (place in its minimal core-being) consists in only one unremovable element, and for that reason it is something we consider in spatial terms when we come to describe it. Belonging to a spatial matrix, a place provides location even if it cannot be reduced to location alone. To be considered temporal, it would have to be constituted by several elements between which a dynamic development emerges. (This is why Bergson insists on the successive character of duration in *Les données*

[2] On this debate, see Casey (1997), especially chapters five and ten.

immédiates de la conscience.) For this reason, place is not essentially an event – even if it belongs to the full description of an event.

Further, if place is indispensably spatial, situation is just as indispensably temporal. The historicity of place adds a temporal dimension to place itself, but it is just that – a *dimension*, a layer, a stratum. Important as it may be, it is not part of its elemental constitution, its minimal being as place: that which is sine qua non for a place to be a place. In contrast, the core-being of situation is characterized by "eventmentality". Something *happens* in a situation. It is not just an assemblage of parts such as those I just enumerated in the case of the recital hall. It is their happening together – *in concert* as it were. ("Con-cert" means literally agreement, union, harmony.)

A situation has several distinguishable basic components, and the relationship between these components is temporal. Each of these components has its own genesis, and together they co-constitute the history of that situation. When these components are concatenated in temporal terms, an *event* emerges: in my example, the event of a musical concert happening in a recital hall. If a place is, minimally, part of space, a situation is in and by itself an event. Being an event, a situation allows for the concatenation of multiple elements in a more or less coherent pattern – coherent enough to allow us to designate it as "*a* situation" or "*this* situation."

3 Situation Filled Out

Despite their apparent affinity with each other, place and situation diverge in the various ways I have been outlining in this brief essay. In this final section, I shall emphasize several ways by which situation takes us in still other directions than does place: it does so insofar as it incorporates synthesis, imagination, and freedom into its "eventmental" emergence.

(a) *synthesis*. A situation is a synthetic compound of diverse elements: not a mere set or sum of such elements but their being held together in at least a momentary unity. This unity is not fixed or steady: as an event in time, a situation is always changing. But at any given moment, it is subject to a synthesis of its parts: say, when the pianist begins to play the first chord of Rachmaninoff's Second Piano Concerto. At the moment, in that event, the pianist, her hands, the piano, the attentive listeners, and all members of the setting (stage, curtains, chairs) are held in a single embrace – these together constituting *that* situation. Minutes later, a different situation obtains even if the elements remain the same: now the pianist is pausing between movements of the concerto, members of the audience are catching their breath or looking at the program notes. Time has moved on, and a distinguishably different synthesis of situational elements has emerged. In short, all the elemental components, though identifiably the same as physical things, realize an altered relationship: each is *situated* somewhat differently with respect to all the other factors. The differences here at stake may not be conspicuous, but in a situational complex any difference, however slight, results in a different synthesis in and of the entire situation.

(b) *imagination*. Because the constituent elements are not all equally easily or well perceived – no one of them takes in *all* of the scene at any given moment – to be composed as a synthetic whole in the manner just discussed requires the infusion of imagination. The facticity of the situation does not suffice to hold it together in one experiential 'Gestalt'. Something else, some other power, must link parts of the scene that are not currently being directly perceived or known. If I am in the concert hall on the right side, and the piano is positioned toward the left, I will not be able to see the pianist's hands. To invoke these ghost hands even as I hear them requires imagination: not necessarily by producing an explicit image of them, but by an imaginative sense of *some hands playing*, no matter how dimly I may be aware of this act of imagination. Such imagination need not be visual only; it can be audial, as when I interpolate notes into a sequence of actually played notes, amplifying the musical mass. Every perceptual synthesis is incomplete and lacunary, and for this reason imagination in its various guises is called upon to fill out the non-plenary presence of what I perceive.

(c) *freedom*. Sartre said famously that "there is freedom only in a situation, and there is a situation only through freedom."[3] (Sartre 1966, p. 629) In his preferred terminology, every situation is an ambiguous combination of freedom with the in-itself.[4] We can just as well say, employing the epicenters of this essay, that every situation involves the factor of place (and thus also of body) and a special freedom that is realized in and through the interaction of the various factors of a given situation. These factors are not only multiple and thus call for imaginative synthesis, but they leave an intrinsic margin of open-ended outcomes. A given situation, taken at any one moment, can turn out in several ways. In part, this is due to the manner in which the elements come to cohere – how they relate to each other – but in larger measure it has to do with how human beings come to configure the same elements in keeping with their own interests and desires. We are not just another factor in a certain situation but one that plays a critical role in animating the situation differentially. If we can be comparatively passive when it comes to place – which is for the most part given as *this place we are now in*: we receive the place even as it receives us – we are able to infuse a factor of freedom into an ongoing situation. This means that the disposition, and the eventual effect, of any particular situation is open and to be determined in quite various ways. The concert pianist might decide to storm off the stage if the audience is rude; or members of the same audience may leave if bored by the same concert. Such differential outcomes manifest the role of freedom in relation to a situation, whose several elements are subject to reconfiguration at any time. If this were not an essential possibility, we would not

[3] Sartre italicizes the first use of "situation."

[4] See ibid., p. 627: "situation is the common product of the contingency of the in-itself and of freedom… [it is] an ambiguous phenomenon."

have a situation, nor would we be an integral part of it; we would instead be confronted with a bare and empty place.

Situations, then, are synthetic wholes in part that are filled out by imaginative acts of those who are part of them, and they are freely determined in their shape thanks to open choices on the part of those beings who are part of these situations in such a way as to alter their configuration at any given moment.

***

Despite the three disparate directionalities I have just discussed, situation does not take us out of place altogether. It takes place and amplifies it into the ambience of a circumstance with multiple characteristics and indeterminate eventualities. It brings place into event by making it part of an interactive whole that at once requires a place in which to occur yet exceeds that place thanks to the outreach realized by situation in its free and imaginative synthesis of multiple constitutive items. Place is not a bare or mute given, and it is much more than a container and more even than a location; it has its own diverse dimensions; but it does not, in and by itself, constitute a situation, which exceeds place even as it requires it as a material condition of possibility. Place calls for situation to extend it into the realm of eventful freedom.

References

Aristotle. 1983. *Aristotle's Physics* Books III and IV. Trans. E. Hussey. Oxford: Oxford University Press.

Casey, E.S. 1997. *The Fate of Place: A Philosophical History*. Berkeley: University of California Press.

———. 2009. *Getting Back into Place: Toward a Renewed Understanding of the Place-World*. 2nd ed. Bloomington: Indiana University Press.

Sartre, J-P. 1966. *Being and Nothingness*. Trans. Hazel Barnes. New York: Washington Square Press.

Place and Placedness

Jeff Malpas

Abstract This paper explores the difference between the notions of place and placedness. This difference relates to an important point of differentiation between genuinely a topographical approach and those other approaches that tend to dominate in the existing literature, including approaches associated with 'situated cognition'. If place is taken as the primary concept, as I argue it should be taken, then that means that being-placed, as it might be viewed as determinative of experience and cognition, has first to be understood in relation to place.

Keywords Being-placed · Bound · Limit · Place · Placedness · Philosophical topography · Philosophical topology · Situated cognition · Situation · Situatedness · Subjectivism

Colin McCahon, *Takaka: night and day*, 1948 (Auckland Art Gallery Toi o Tāmaki, NZ) – in McCahon's words: "landscape as a symbol of place and also of the human condition". (McCahon (1972, p. 19). By kind permission of the Colin McCahon Research and Publication Trust)

J. Malpas (✉)
University of Tasmania, Hobart, TAS, Australia
e-mail: Jeff.Malpas@utas.edu.au

T. Hünefeldt, A. Schlitte (eds.), *Situatedness and Place*, Contributions To Phenomenology 95, https://doi.org/10.1007/978-3-319-92937-8_3

1. Is there a philosophically significant difference between the notions of *place* and of *being-placed* – of what might be termed *placedness*? The question might also be put in terms more directly relevant to the idea of situated cognition by asking whether there is a philosophically significant difference between *situation* and *being-situated* or *situatedness* (assuming 'situation' is understood, *prima facie*, as that *wherein* one is situated). That there is a difference is easily elided by the fact that, in the latter case, the notion of situation is frequently discussed alongside and often interchangeably with being-situated. One might thus talk equally of one's situation or one's being-situated without any necessary shift in meaning.

The question might seem to be a relatively minor one, but it relates to an important point of difference between a genuinely topographical or topological mode of thinking,[1] the sort of thinking that both Ed Casey and I have tried to develop, each in our own way,[2] and certain modes of thinking that may draw upon elements of the topographic, but for which *topos*, place, is actually a secondary concept. My suggestion here will be that the focus on being-placed, placedness, or, as it may also be put, on being-situated, situatedness, can itself obscure the question of place, and that the question of place must come before any question of being-placed or placedness – even though it is only through being placed that we gain access to place. The difference may also be important in marking out a further and more specific difference between the way in which notions of place and situation enter into much cognitive scientific discourse, including that of situated cognition, and the way the notion of place, especially, may appears within broader forms of place-oriented discourse.

2. That there is a *prima facie* difference between place and placedness seems undeniable – at least if one gives a little thought to the matter. In simple terms, 'placedness' or 'being placed' names a characteristic, even if generalizable, of that which is placed, whereas 'place' names that to which what is placed stands in relation. *Placedness* would thus seem, on the face of it, to presuppose *place*. On that basis, there can be no placedness without place, and the two notions are inextricably bound together even though they are also distinct – the same reasoning may also be applied to the notions of situation and situatedness or being-situated. Yet what appears to be a simple and obvious difference here conceals a larger set of complications. There is a general tendency for place and placedness not to be distinguished even in discussions in which the concepts play an important role – the most obvious indication of which is the widespread identification of place with some notion of *meaningful space,* that is, with space as it is given meaning by a subject. Such a way of thinking about place is evident, for instance, in the work of one of the most influential writers on place and space, Yi-Fu Tuan, who writes that "in experience, the meaning of space often merges with that of place. 'Space' is more abstract than 'place'. What begins as undifferentiated space becomes place as we get to know it

[1] 'Topographical' and 'topological' are here used, as I have deployed them elsewhere, more or less as synonyms – as I would see it, one emphasises the 'writing' of place and the other its 'saying'.
[2] See, for instance, Malpas (2018), and Casey (2009).

better and endow it with value" (Tuan 2001, p. 6) Here it seems that space and place essentially exist on a continuum in which the move towards place is also a move toward the human valuation of space. Tuan thus distinguishes place from space, after a fashion, but the way he does this effectively reduces place to a variety of space – space as given human valuation – and since he treats place as dependent on a human mode of apprehending space, he leaves no room for place other than as tied to such apprehension. Place is apprehended space, or as we might also say, it is space understood as it belongs to our being in space and our responding to it – it is, one might say, the space of our situatedness and understood in terms our situatedness or placedness.

The tendency to treat place in this way reflects a broader lack of attentiveness to place as a genuinely *sui generis* concept (even among many of those who seem otherwise to take place as a significant notion) – a lack of attentiveness that, whatever else it signifies, often amounts to an effective reduction of place to placedness or the replacement of place by placedness. When this happens, the very notion of place undergoes an important shift, since placedness no longer involves standing in a genuine relation *to place*, but instead seems to imply that place somehow belongs to the character of that which is placed – as the valuation of space arises on the basis of the human being in space. If this sounds odd or obscure – and it ought to – then the reason is simply that it is so. Moreover, even though this implied shift to placedness over place is commonplace, its oddity or obscurity typically goes unremarked because the shift itself is seldom acknowledged.

To illustrate what is at issue here let me take as an example an idea developed by John Campbell.[3] In his *Past, Space and Self*, Campbell argues, primarily against the Strawsonian claim that subjectivity requires objectivity (or at least that a subject requires objects in order to operate as a subject), that one could conceive of a case in which an agent was capable of moving itself in a coordinated fashion and yet has no sense of space other than as purely subjective. What Campbell apparently has in mind is a case in which an agent guides its movements according to subjectively presented features – something that he suggests can be illustrated in the case of human agency by navigational instructions of the sort: "steer always with the wind at your back" or "keep on a course that has the setting sun at your right shoulder" (Campbell 1995, p. 14). Important to Campbell's account is the idea that the space he has in mind here, or at least the grasp of that space, is dynamic – it is essentially tied, not to some static model of space, but directly to action – Campbell talks in fact of this as a mode of "egocentric space" that is "immediately used by the subject in directing action" (Campbell 1995, p. 14). Here egocentric space seems to be in some sense a structure of the acting subject rather than referring to something that stands apart from the subject.

The idea of an *egocentric* space such as Campbell describes that is independent of any notion of space as *objective* seems exactly analogous to the idea of a mode of

[3]This is an example to which I also make reference to in *Place and Experience* – see Malpas (2018), pp. 135–136. There my concern is with the nature of spatial understanding and the interdependence of a conceptual grasp of space with a grasp of objects.

placedness that is distinct from place. One might argue about whether Campbell's account, as he characterizes it, is fully consistent or coherent – the maritime examples he uses are not entirely convincing, and it may be that the sort of system he has in mind is actually much better characterized in purely functionalist terms (the aligning of certain perceptual inputs with behavioral outputs) that need not imply any necessary reference even to spatiality. Still, Campbell's example does seem to provide us with an example of a position that allows a reasonably clear separation of what might be interested as a mode of placedness as apart from a notion of place. Notice that Campbell's account really only works for the analysis of individual behavior – it is directed at an analysis of a mode of subjectivity and so it is perhaps not surprising that it involves a notion of a subjective space or being-place. Campbell does not reject the idea of a notion of place that goes beyond such subjectivity, just as he allows a notion of objective space that goes beyond egocentric space. The point of disagreement one might have with Campbell is to what extent the idea of a subjective space is genuinely independent of (even if not reducible to) a notion of intersubjective or objective space – and so also whether the being-place on which he focuses can be made sense of apart from place. Campbell tends to view his concept of egocentric space as indeed independent.

Campbell's discussion might be taken to suggest that the difference between place and placedness is the same as or at least analogous to the distinction between egocentric or subjective space and objective space – a suggestion that would, however, set place in a quite opposite position to that which is common in much of the literature according to which place is most often taken to be associated with the subjective rather than the objective. Yet although the shift away from a subjective understanding of place is important, we ought to resist the idea that placedness is to be identified with *space as subjective* and place with *space as objective*. Part of the reason for this, of course, is that place *and space* are distinct notions, even though related, but more important is the fact that place is not to be construed as an objective structure to be set against the subjective. If one takes subjective and objective to be correlative notions or structures, as in one sense they surely are, then subjective and objective only appear in relation to one another and within a larger frame that encompasses both. Such an encompassing frame can belong wholly neither to subject nor object. This seems to me a point well-illustrated by Donald Davidson's thinking around the notion of triangulation – itself an essentially topographical or topological notion.[4] Davidson treats, not only subjective and objective, but also the intersubjective, as part of a single interrelated structure articulated through the idea of triangulation which here names both an epistemological and ontological formation. The triangular structure within which subjective, intersubjective and objective are worked out can be taken to be equivalent to the notion of place as understood not merely as some location in the world, but rather as that within which any sort of appearance or encounter is possible.

---

[4] For more on triangulation and topology see my "Self, Other, Thing: Triangulation and Topography in Post-Kantian Philosophy" (Malpas 2015a).

To go back to Campbell's example, however, one could read the sort of case Campbell presents, not merely as attempting to establish the idea of an egocentric or subjective space as an independent or sui generis notion that is nevertheless part of a larger framework of spatial and topographic elements, but instead as showing that there is no necessity for anything beyond a notion of egocentric or 'subjective' space in understanding the possibility even of agency that seems to involve spatial orientation and direction. Since what is required for spatial agency is a way in which the spatial engages with the agent's capacities for action, then all that is needed is a subjectively presented space (which need not imply a subjectively represented space) – and any space must be subjectively presented if it is to engage with action. Indeed, one might argue that all action is action in a subjectively presented space – or, as one might also put it, in a subjectively presented environment. One might put this point more generally and say that, on this reading, neither action nor cognition need involve anything other than the direct responsive interaction of an acting subject with its environment. Moreover, on this account, there need be no notion of an internal representation of the environment, but neither need there be any idea of an environment that stands apart from, or that can be characterized apart from, the acting subject. This means that such an account is compatible both with what might be thought of as traditional 'idealist' or 'subjectivist' positions and with positions that are 'realist' or 'physicalist'.

3. The sort of account at issue here – the sort of account that might be drawn out of Campbell's position – is not unlike that which one can find in some accounts of situated or embodied cognition. One can characterize such accounts as subjectivist, since they rely on a mode of subjective presentation, one can take them as objectivist, in the same sense that behaviorism is objectivist, or one can take them, as they often do, as standing outside of the subject-object distinction altogether – which is how they often present themselves. A useful example of this latter sort of approach is Hubert Dreyfus' notion of embodied coping as developed in many publications over the last 40 years or so, since the first publication of his ground-breaking book *What Computers Can't Do* (Dreyfus 1972).[5] Dreyfus takes our being in the world to be determined in terms of our activity, and without any representational intermediaries. In the terms I have used, however, there is no 'place', in Dreyfus, that is distinct from 'placedness'. Instead, there is only the causal-physical structure of the world and the direct interactive responsiveness with the world that belongs to the coping agent. Dreyfus' account thus combines a phenomenology grounded in the early Merleau-Ponty with a behaviorism largely derived from (or at least convergent with) key aspects of the work of Gilbert Ryle. Just as Dreyfus seems not to distinguish place from placedness, so neither does situation appear apart from situatedness or being-situated. 'Situation' becomes simply the particular differential orientation in the world that belongs to the acting subject and that is necessarily implied by the subject's capacity for engaged coping.

Dreyfus is an interesting figure to consider in this context, since it is Dreyfus who has largely been responsible for the introduction of phenomenological, and

[5] A more recent edition appeared as *What Computers Still Can't Do* (Dreyfus 1992).

especially Heideggerian, influences into contemporary cognitive science. This has been particularly so in respect of the anti-representationalist – one might even say 'anti-cognitivist' – tendency that is associated with both embodied and situated cognition. As elaborated by Dreyfus – most notably in his *Being-in-the-World* (Dreyfus 1990) – Heidegger shows us that our primary mode of being-in-the-world is given in terms of action rather than knowledge, and on the basis of our engaged involvement rather than our detached observation – on the basis of *praxis* rather than *theoria*. Yet although this has been a key element in Dreyfus' appropriation of Heidegger into cognitive science, as well as of his reading of Heidegger more generally, it is a highly problematic reading of Heidegger, and the problems associated with it are not far distant from the issues relating to the distinction I have suggested between place and placedness.

Although it is true that Heidegger rejects the claim that we can understand being-in-the-world on the model of a detached, 'scientific' understanding, this does not mean that the standpoint of the 'theoretical' is thereby taken to be essentially secondary to the 'practical'.[6] What Heidegger is concerned to reject is the prioritization of the scientific projection of the world that he actually takes to constitute a form of subjectivism and nihilism, and which constitutes only a particular development of a certain mode of theory. Our engaged involvement in the world is an involvement that can take both practical and theoretical forms, with theory itself having its own mode of *praxis*. That the theoretical cannot be taken as secondary is especially obvious once one reflects on the role of philosophy, of thinking, in Heidegger's account – and, in the later work, such thinking is fundamentally about a certain sort of attunement to place, even a mode of contemplation of place and our relation to it.[7] The latter itself depends on distinguishing place from our own being-in-place, and only if we do indeed distinguish place from being-in-place, from placedness, can we make sense of the Heidegger's topological project (whether in *Being and Time* or elsewhere) as also a genuinely ontological project that is addressed to *being* rather than merely to *human* being, and so as a project that is not to collapse into some form of subjectivism – which is what the prioritization of the 'scientific' itself tends towards.

A central problem that afflicts Heidegger's account in *Being and Time* is that placedness or situatedness readily appears there as a structure of Dasein and Dasein is itself understood as identical with the essential structure of human being. Already one can see the dangers of an incipient subjectivism here – even though such subjectivism is one of the things Heidegger aimed, in *Being and Time*, to overcome. Dasein is characterized by Heidegger as 'being-the-world', and although this does indeed shift the focus away from an internalized form of subjectivity that is set against the world, it nevertheless also runs the risk of effectively subjectifying the world, since the being-in that belongs to being-in-the-world is itself grounded in Dasein's own projection of possibilities (that projection being precisely the projection of world). The position is complicated, of course, by the fact that Dasein here

[6] See my discussion in *Heidegger's Topology* (Malpas 2006), pp. 140–141.

[7] See Malpas (2006), *Heidegger's Topology*, Chapt. 5.

does not name a mode of being-in that is apart from being-with or being-alongside (and so the subjectivising tendency at work here is by no means unequivocal or unambiguous), but it is a tendency that even Heidegger himself acknowledged.[8] In *Being and Time*, the main focus tends to be more on what is effectively a mode of placedness than it is a mode of place – and this itself reflects the fact that *Being and Time* is lacking in any explicitly topological vocabulary (certainly in comparison with the later thinking), and the notion of existential spatiality that is set out in the early part of the work (and which is, in any case, said to be secondary to temporarily) is actually closer to a mode of placedness than of place.[9]

It is out of Heidegger's recognition of the problems that remain within *Being and Time* – problems that can be seen centrally to rest on the work's treatment of space and time, as well as place – and so out of his attempts to resolve those problems, that Heidegger's thinking undergoes a significant shift. In simple terms, the shift at issue here is from a position in which *place* is a projection of human being (or better, of Dasein as the essence of human being) to one in which *human being* is a projection or 'function' of place (and so human being comes to belong essentially to place). This shift is thus one that can be characterized as being from *place* as the projection to *placedness* as that which is projected, but, at the same time, what also occurs is a separating out of place from placedness and the emergence of a genuinely *sui generis* concept of place – the latter occurring largely through Heidegger's engagement with Hölderlin beginning in the mid-1930s.[10] The shift to place in Heidegger's thinking, which is to say the development of his thinking as having the form of what he calls a 'topology', also brings with it a rethinking of Dasein itself, since Dasein now names place rather than any form of placedness – it names the being of place (and the place of being) rather than being-in-place or as we might also put it, situatedness.

When one takes seriously the topological character of Heidegger's thinking as that has been briefly sketched here, then it soon becomes evident that Heidegger's thought diverges significantly from much of what is taken for granted within

---

[8] See, for instance, Heidegger's comment in *Contributions* that "In *Being and Time* Da-sein still has an appearance that is 'anthropological,' 'subjectivistic,' 'individualist,' etc." (Heidegger 2012, p. 233). Heidegger makes this comment still emphasising that *Being and Time* itself aimed to take issue with such anthropologism and subjectivism, and yet part of the difficulty is this remains an issue as Heidegger's own focus on the matter here and elsewhere itself suggests.

[9] Richard Schacht (2013, pp. 1–24) focuses on what he terms the 'topology' of Heidegger's early work, and especially *Being and Time*, and briefly takes issue with my own emphasis (which he takes to be characteristic of most commentaries) on the topology that becomes explicit in the later thinking (Schacht 2013, p. 21, n.19). I do not disagree with the claim that topology runs through all of Heidegger's work, and this is a key claim in *Heidegger's Topology*, but what I think Schacht overlooks here is precisely the way *Being and Time* effectively neglects place in favour of placedness at the same time as it also favours place (inasmuch as it is addressed) as indeed a 'projection' rather than as projecting.

[10] See e.g. Julian Young, *Heidegger's Philosophy of Art* (Young 2004). See also my discussion in "Die Wende zum Ort und die Wiedergewinnung des Menschen: Heideggers Kritik des 'Humanismus'" ["The Turn to Place and the Retrieval of the Human: Heidegger's Critique of 'Humanism'"] (Malpas 2017).

contemporary cognitive scientific thinking. Indeed, I would argue that Heidegger cannot be assimilated to a cognitive scientific perspective without significant distortion of his thinking. This ought already to be obvious, however, from the fact that his own focus is on *being* rather than on the structures of human *cognition* (which is why Heidegger denies his position is anthropological or humanistic) or, at least, the latter is significant for Heidegger only inasmuch as it sheds light on the former (and it will do so only if the question of human cognition is already taken to lead on to the ontological question).

In this respect, Heidegger's position can be contrasted with that of Merleau-Ponty, whose early work is more directly and readily assimilable to a cognitive scientific perspective (which is why Dreyfus' account actually tends to be much closer to the French thinker than the German), but which does not bring the question of being to the fore. Indeed, in contemporary cognitive science, and in many contemporary fields in which place seems to figure, it is Merleau-Ponty who most often occupies center stage, and this is surely because Merleau-Ponty, especially in his earlier work, offers a way of thinking about place that is both more accessible and that tends to treat place much less equivocally, which is to say, in terms that allow its construal as more or less indistinguishable from placedness. One indication of this is the early Merleau-Ponty's tendency to emphasis *the body* rather than place, in spite of the fact that it is hard to make sense of the body independently of place. This tendency reflects a more widespread tendency to look to the body as some of kind of foundation or *subjectum* – to treat the body as an explanatory ground rather than as itself in need of explanation. What the body *is* cannot be taken for granted and the nature of the body remains always in question just so long as the body is treated as prior to or as apart from a mode of being-in-place (which means that the analysis of *place* has to come before the analysis of the *body*).[11] Unlike Heidegger, Merleau-Ponty is also more congenial to the strongly non-cognitivist approach that is characteristic of much contemporary research that purports to take place as a key theme. The reason for this is largely that the emphasis on the body enables a stronger focus on purely bodily and behavioral responses – which is precisely what one finds in Dreyfus – and which is therefore also more amenable to analyses in terms of underlying bodily processes and structures. This is also, of course, a reason why such an emphasis is indeed more congenial to cognitive scientific approaches and to the increasing prominence of ideas and approaches taken from contemporary neuroscience.

4. Earlier I noted the possibility that, understood as distinct from placedness, place might seem to converge with the idea of objective space or even with the idea of objectivity. I also noted that although it is indeed mistaken to construe place

---

[11] This point is very clear in Heidegger – see my "Heidegger, Space and World" (Malpas 2012), pp. 311–312 – although it also means that he is often treated as neglectful of the body (especially in his treatment in *Being and Time*). It is worth noting that our being-in-place is not a function of our being-embodied, but rather, our being-embodied is itself derivative of our being-in-place – in exactly the same way that, as Heidegger points out in his 1929 *Kantbuch*, our dependence on the senses is a function of our finitude rather than our finitude being a consequence of our dependence on the senses – see *Kant and the Problem of Metaphysics* (Heidegger 1997, pp. 18–19).

entirely subjectively, it is nevertheless inappropriate to regard place as therefore to be construed in solely objective terms either. Precisely because place *encompasses* both the subjective and the objective, thinking in terms of place is amenable to thinking in terms of both subjective *and* objective, and entails the irreducibility and indispensability of both of these even though they are nevertheless also inevitably entangled with one another.[12] Subjectivity and objectivity are both structures that appear only within or in relation to place and arise only out of the engagement with and in place.

In analogous fashion, one cannot simply identify a topographical or topological mode of thinking with either a first-personal or a third-personal approach, as if place were just one or the other, but will always involve the interplay between them. The first-personal and the third-personal are thus both to be understood only from within the framework of place. In this way the topographical thereby also has to be understood as standing outside of the usual contrast between the phenomenological and the empirical scientific – which are themselves often identified with the first personal and the third-personal. Adopting a properly topographic approach does not mean ruling out such perspectives, but it does mean recognizing their location within a larger landscape that allows of multiple descriptions. Moreover, while it allows that there will be relations between different sorts of descriptions here, topographic thinking nevertheless refuses to allow any unambiguous reduction between descriptions or any determinate level of description that underpins all description.[13]

Such an approach to the notions of the subjective and objective (and the intersubjective) as well as the first-personal and third-personal, along with the emphasis on descriptive indeterminacy and multiplicity, are indeed characteristic features of the work of those thinkers who I would argue exemplify a mode of genuinely place-oriented thinking: thinkers such Davidson, but also Heidegger and even Gadamer – those who also exemplify what I have elsewhere characterized as a mode of hermeneutic-topographical thinking.[14] The emphasis on the hermeneutical here (in contrast, notably, with the phenomenological) is also important: understanding is grounded in the placedness or being-placed of the one who understands just as that being-placed is itself determined by the general structure of place as well as the singular character of that very place that is at issue (and so the singular character of that being-placed).[15] Moreover, the hermeneutical also brings with it a tendency to treat subjectivity and objectivity, first person and third person, as both structures

[12] This is essentially the line of approach I adopt in *Place and Experience*. There I argue that "place is not to be viewed as a purely 'objective' concept … [that is] a concept to be explicated by reference to objects existing in a physical space … [and] neither is it a purely 'subjective' notion … both subject and object are … 'placed' within the same structure, rather than one or the other being the underlying ground for that structure" (Malpas 2018, pp. 33 & 35).

[13] See "Self, Other, Thing: Triangulation and Topography in Post-Kantian Philosophy" (Malpas a).

[14] See my "Placing Understanding/Understanding Place" (Malpas 2016).

[15] On the singularity of place see my "Place and Singularity" (Malpas 2015b).

that are part of a larger 'event' or, as I would say, a larger 'taking place'. In this way, hermeneutics can be seen to bring a mode of topography or topology along with it even as topography draws us into proximity with the hermeneutical.[16]

Even though there may be reasons for taking the hermeneutical to stand in a particularly close relation to the topographic or topological, still the emphasis on place does not mean ruling out either phenomenological or empirical scientific approaches just as it does not imply of rejecting notions of subjectivity or objectivity or of ignoring either first personal or third personal perspectives. Instead, the emphasis on place allows us to attend to all of these without giving absolute priority to any one. This means too that the emphasis on place as distinct from placedness or being-placed does not entail the dismissal of placedness, in particular, as a significant notion. Indeed, it is only if one retains a clear sense of the distinction of place from placedness that either of these concepts can properly remain in view. The tendency to ignore place actually results, not in the prioritization of placedness, but rather in the obliteration of both place *and* placedness, since the latter itself depends on the former. This is why Heidegger's *Being and Time* serves as an important opening up of the way into the thinking of place since the manner in which it takes up the idea of situatedness is such as already to invoke a notion of place even though it is also a notion that it does not, in the end, properly address – hence the incomplete and uncompletable character of the work.

If we distinguish place from placedness, thereby also retaining both concepts even as we also insist on the primacy of place itself, then perhaps we must also distinguish between two notions of placedness – although one might argue that only one of these is properly so called.[17] The first of these forms of placedness is the placedness that stands in an essential relation to place. This is the placedness that just because it is indeed a being *in-place* thereby calls upon a notion of place that is nevertheless also distinct from it. The other notion of 'placedness', and it is here placed in quotation marks to indicate its anomalous character, is something like Campbell's notion of a purely egocentric or subjective 'space' or perhaps as we may also term it a purely behavioral 'space'. This notion of 'placedness' is only ambiguously characterized as indeed a mode of space or spatiality or as a mode of placedness. This is because the 'placedness' at issue here is really an attribute or set of attributes belonging to a creature, agent, or *subject* – what might be termed a mode of behavioral responsiveness that could, as I noted above, be characterized purely functionally. It is certainly unclear to what extent such a notion of 'placedness'

[16] The way the hermeneutical appears here draws attention to another important feature of the topographic – one evident in the work of just those thinkers I have already invoked, especially Davidson and Heidegger – namely the connection between place and language. This may be thought somewhat odd given my emphasis on the distinction of place from being-placed and so the insistence on place as a structure that comes before the subjective. Yet as place is not to be understood as primarily subjective, neither, I would argue, is language, though this point depends on distinguishing the specificity of speech and script form the very articulation of the world that these make salient and express.

[17] On the assumption that situatedness is understood as distinct from situation, then the notion of situatedness will present a similar equivocity.

requires a notion of *place* in its explication or even a notion of *space* (other than as a nexus of causal relatedness).

Place and placedness disappear in the face of the reduction of being-placed to a property of the agent; but place and placedness, and so the distinction between them, also disappear in the fact of the widespread tendency to treat place (and so placedness along with it) as a product of the subject or of the interaction between subjects, in other words, as subjective or intersubjective (psychological, social, cultural, or political) *constructions*. On this account, there may still be a notional distinction of place from placedness, but since there is no *sui generis* notion of place or placedness, both being mere 'effects' of supposedly more basic structures and processes, so the distinction turns out to be merely notional. Similarly, one cannot treat place as having a distinct character apart from placedness or being-placed, since there is no place that is not itself a psychological, social, cultural or political structure, process, or phenomena, and no being-placed that is not such either.[18]

Holding fast to the distinction between place and placedness means holding both to the idea of placedness or being-placed as involving a genuine relation *to place* and a refusal of any reduction of place to something else or its treatment as merely derivative. This means according a particular *ontological* status to place, and, indeed, place itself already stands in a very particular relation to *being*. To be, one might say, is to be placed – and this idea is one that Aristotle invokes when he repeats, in *Physics* IV, the Archytyan dictum that to be is to be somewhere.[19] This means that for any being, what it is for it to be is for it to be in place – to be placed – and this opens up the question (as it is opened up, if also somewhat problematically, in Aristotle) as to what place itself might be.

The question about place thus emerges through a question about the being-placed of some thing, and yet the former question is indeed a question about place and not merely about being-placed. The question about the being of place is a peculiar one, however, since it cannot properly be a question that asks after the being of place as if this concerned merely the being of this or that place or as if this concerned some independent mode of being that might or might not attach to place. If to be is to be in place, then being and place appear as correlative notions, so closely tied together that they can barely be separated. If this is often overlooked, the reason is simply that all too often we treat place as identical with places – with individual *locations* or *locales*. Yet to ask after what place is, and so after the mode of being of place, we are really asking after what place is independently of any individual place, independently of any specific location or locale. This question is almost indistinguishable form the question of being, and yet, in recognizing this, the question of being itself appears in a new light, as essentially topographical or topological – hence Heidegger's claim that his own inquiry into being takes the form of a *topology*.[20]

[18] For more on constructionist construals of place see my "Thinking topographically: place, space, and geography" (Malpas 2017).

[19] Aristotle, *Physics* IV, 208a30.

[20] See Heidegger "Seminar in Le Thor 1969" (Heidegger 2004, p. 41 & p. 47).

Recognizing the way the question of place and the question of being converge also allows us to see how the question of being that is now seen to be at issue here is not a question that concerns the being of this or that – it does not concern the being merely of a being or of beings any more than the question of place at issue concern the 'place' or 'being-placed' of any thing or things. Instead what is at issue is indeed the being that belongs first *with place* in the same way as the place at issue is the place that belongs first *with being*.

5. The question of place, though it is indeed only to be approached through our own place, and so though our own placedness, is a question that goes beyond ourselves, beyond even those other selves and other things around us, and that thereby encompasses that wherein we always already find ourselves – a 'wherein' that points in the direction of the world, and yet also indicates the way the world itself begins only in and through place. Thinking, no matter where it eventually arrives, begins only in and out of this *being-placed* which is always a being *in relation to place.* It is thus that the question at issue here is indeed a question that concerns more than just our own *being* or our *own place.* Place arises as a question out of being placed, out of placedness, but it certainly does not remain as a question merely of placedness. It is precisely because the question of placedness opens out into the question of place in this way that the question of placedness forces us to attend to our own radical finitude, our own boundedness, our own limit, and so, thereby, our own being – which is given only in and through this limit. Not only is place a notion that is itself tied to the idea of limit (although the limit that belongs to place must be understood as enabling rather than simply constraining),[21] but in recognizing the placed character of our own being we are also forced to recognize the way in which the being that is proper to us – a being that is a being in place – is also a being in which we are opened up to place and so to being. We thus come back to our own being through the encounter with the placed character of our being (through our being here/there) in which we are also opened up to place itself. It is in precisely this direction (the direction of a genuine topology or topography) that the thought of the later Heidegger moves, but it is a direction that can also be seen as indicated in the work of a host of thinkers, writers, and artists – in the work of any who attend to the real manner in which the world happens, the real manner in which the world does indeed 'take place'. In such work placedness and place appear together, though never as simply conflated. Thus a painter like the New Zealander, Colin McCahon, almost all of whose work can be said to explore both the human being in place and the place that is thereby revealed (a place that in his work also opens up to the sacred), can be said to be a painter of place no less than Heidegger is a thinker of place, even though McCahon's work is also, of necessity, like Heidegger's own thinking, an exploration that occurs only in and through the human experience of place. The point, of course, is that the experience is never an experience of itself alone.

[21] On place and limit, see, for instance, the discussion in "Thinking topographically: place, space, and geography" (Malpas 2017). The topic also arises in *Heidegger's Topology* and elsewhere. Heidegger famously emphasises that a limit or boundary is "not that at which something stops but… that from which something begins its presencing" (Heidegger 1971, p. 154).

References

Campbell, J. 1995. *Past, Space and Self.* Cambridge: MIT Press.
Casey, E. 2009. *Getting Back into Place*. 2nd ed. Bloomington: Indiana University Press.
Dreyfus, H.L. 1972. *What Computers Can't Do*. Cambridge: MIT Press.
———. 1990. *Being-in-the-World: A Commentary on Heidegger's Being and Time, Division I.* Cambridge: MIT Press.
———. 1992. *What computers can't do*, 2nd edn. Cambridge: MIT Press.
Heidegger, M. 1971. Building dwelling thinking. In *Poetry, Language, Thought*, ed. A. Hofstadter. New York: Harper & Row.
———. 1997. *Kant and the Problem of Metaphysics,* 5th enlarged edn. Trans. R. Taft. Bloomington: Indiana University Press.
———. 2004. *Four Seminars.* Trans. A. Mitchell and F. Raffoul. Bloomington: Indiana University Press.
———. 2012. *Contributions to Philosophy.* Trans. R. Rojcevicz and D. Vallega-Neu. Bloomington: Indiana University Press.
Malpas, J. 2018. *Place and experience*, 2nd edn. Abingdon: Routledge.
———. 2006. *Heidegger's Topology*. Cambridge: MIT Press.
———. 2012. Heidegger, space and world. In *Heidegger and Cognitive Science*, ed. J. Kiverstein and M. Wheeler. London: Palgrave Macmillan.
———. 2015a. Self, other, thing: Triangulation and topography in post-Kantian philosophy. *Philosophy Today* 59: 103–126.
———. 2015b. Place and singularity. In *The intelligence of place: Topographies and poetics*, ed. J. Malpas, 65–92. London: Bloomsbury.
———. 2016. Placing understanding/understanding place. *Sophia*. https://doi.org/10.1007/s11841-016-0546-9.
———. 2017. Die Wende zum Ort und die Wiedergewinnung des Menschen: Heideggers Kritik des "Humanismus" [The Turn to Place and the Retrieval of the Human: Heidegger's Critique of "Humanism"]. In *Heideggers Weg in die Moderne. Eine Verortung der "Schwarzen Hefte"* [Heidegger Forum 13], eds. H.-H. Gander und M. Striet. Frankfurt a.M: Klostermann.
Malpas, Jeff. 2017. Thinking topographically: Place, space, and geography. *Il Cannocchiale: rivista di studi filosofici* 42: 25-54.
McCahon, C. 1972. *Colin McCahon: A Survey Exhibition*. Auckland: Auckland City Art Gallery.
Schacht, R. 2013. The place of mimesis and the apocalyptic: Toward a topology of the near and far. *Contagion: Journal of Violence, Mimesis, and Culture* 20: 1–24.
Tuan, Y.-F. 2001. *Space and Place The Perspective of Experience*. 5th ed. Minneapolis: University of Minnesota Press.
Young, J. 2004. *Heidegger's Philosophy of Art*. Cambridge: Cambridge University Press.

Merleau-Ponty, Lived Body, and Place: Toward a Phenomenology of Human Situatedness

David Seamon

Abstract In this chapter, I draw on French phenomenologist Maurice Merleau-Ponty's understanding of perception and corporeal sensibility to consider the significance of human situatedness as expressed via *place* and *place experience*. To illustrate how Merleau-Ponty's conceptual understanding might be applied to real-world place experiences, I draw on two sources of experiential evidence, the first of which is a passage from Colombian writer Gabriel García Márquez's magical-realist novel, *One Hundred Years of Solitude* (1967/1970). The second source is a set of first-person observations describing events and situations that I happened to take note of during weekday walks between my home and university office over the course of several months. My aim is to use these two narrative accounts as a means to illustrate, via vignettes of everyday human experience, Merleau-Ponty's central concepts of perception, body-subject, and lived embodiment. I contend that these accounts substantiate Merleau-Ponty's phenomenological claims and point to additional significant elements of human situatedness and place experience.

Keywords Merleau-Ponty · Phenomenology · Place · Situatedness · Body-subject · Body schema · Lived body · Environmental embodiment · Habit · Corporeality

1 Introduction

> [S]pace always precedes itself.... It is of the essence of space to be always "already constituted," and we shall never come to understand it by withdrawing into a worldless perception. We must not wonder why being is orientated, why existence is spatial, [...] and why [our body's] coexistence with the world magnetizes experience and induces a direction in it. The question could be asked only if the facts were fortuitous happenings to a subject and an object indifferent to space, whereas perceptual experience shows that they are presupposed in our primordial encounter with being, and that *being is synonymous with being situated*. (Merleau-Ponty 1962, p. 252 [italics added])

D. Seamon (✉)
Department of Architecture, Kansas State University, Manhattan, KS, USA
e-mail: triad@ksu.edu

T. Hünefeldt, A. Schlitte (eds.), *Situatedness and Place*, Contributions To Phenomenology 95, https://doi.org/10.1007/978-3-319-92937-8_4

In this chapter, I draw on French phenomenologist Maurice Merleau-Ponty's understanding of perception and corporeal sensibility to consider the significance of human situatedness as expressed via *place* and *place experience*.[1] I begin with an overview of phenomenology and a phenomenological perspective on place. I then discuss Merleau-Ponty's phenomenology of perception and lived body, emphasizing themes and concepts relating to people's lived relationship with place. To illustrate how Merleau-Ponty's conceptual understanding might be applied to real-world place experiences, I draw on two sources of experiential evidence, the first of which is a short, 500-word passage from critically-acclaimed Colombian writer Gabriel García Márquez's magical-realist novel, *One Hundred Years of Solitude* (García Márquez 1967/1970, pp. 246–248). The second source is a set of first-person observations describing events and situations that I happened to take note of during weekday walks between my home and university office over the course of several months. The aim is to use these two descriptive sources for grounding, clarifying, and extending Merleau-Ponty's broader philosophical claims, which are often difficult to understand, especially for newcomers.

In doing phenomenological research, one faces the difficult conceptual and methodological question of interpretive accuracy and trustworthiness (Finlay 2011; Seamon 2017; Seamon and Gill 2016; van Manen 2014; Wachterhauser 1996). How, in other words, does the phenomenologist establish a convincing interpretive link between real-world experience and conceptual generalization? In this chapter, I suggest that one methodological means to reduce interpretive error is the use of contrasting but related texts that might offer a multifaceted illumination of the phenomenon being studied – in this case, place, lived embodiment, and human situatedness. Here, I draw on three such texts: first, Merleau-Ponty's *Phenomenology of Perception* (Merleau-Ponty 1962); second, García Márquez's fictional narrative; and, third, the first-person observations of my weekday walks. I call this method *triangulation*, whereby I use the three texts to illuminate, amplify, and validate each other.

In the literature on qualitative research, triangulation is more typically defined as a research approach whereby the researcher draws on multiple methods, data sources, evaluators, and conceptual approaches as a means to identify different lived perspectives and to corroborate evidence from different data sources (Creswell 2007, p. 208; Willig 2001, p. 142; Yardley 2008, pp. 239–240). In this chapter, my use of triangulation is somewhat different in that the three texts I use for interpretive corroboration are of different descriptive "levels" – in other words, one text is Merleau-Ponty's philosophical presentation, and the other two texts are narrative accounts of everyday experiences and events. My aim is to use these two narrative

[1] The starting point for this chapter is an invited paper, "Situated Cognition and the Phenomenology of Place: Lifeworld, Environmental Embodiment, and Immersion-in-World," prepared for a symposium on "Situated Cognition and the Philosophy of Place," organized by "Thomas Hünefeldt and Annika Schlitte, and held at the 6th International Conference on Spatial Cognition," Rome, Italy, September, 2015. Portions of this chapter are based on that invited paper, which was published as Seamon 2015.

accounts as a means to illustrate, via vignettes of everyday human experience, Merleau-Ponty's central concepts of perception, body-subject, and lived embodiment. I contend that these accounts substantiate Merleau-Ponty's phenomenological claims and point to additional significant elements of human situatedness and place experience.

2 Place, Phenomenology, Natural Attitude, and Lifeworld

I define place as *any environmental locus that gathers human experiences, actions, and meanings spatially and temporally* (Seamon 2013, p. 150; Seamon 2018, p. 2). Places range from intimate to regional scale and include such environmental situations as a frequently used park bench, a cherished home, a favored neighborhood, a city associated with fond memories, or a geographical locale that is a regular vacation destination. Experientially, places are multivalent in their constitution and sophisticated in their dynamics. On one hand, places can be appreciated, loved, and cared for; on the other hand, they can be unappreciated, loathed, and despoiled. For the persons and groups involved, a place can invoke a wide range of sustaining, neutral, or debilitating actions, experiences, situations, and meanings (Seamon 2014, 2018).

To better understand the constitution of place and place experience, I draw on a phenomenological approach, which can be defined most simply as the description and interpretation of human experience, particularly its tacit, transparent dimensions usually unnoticed in everyday living.[2] Phenomenologists use the term *natural attitude* to identify the unquestioned way in which people automatically accept the taken-for-grantedness of day-to-day life. In turn, phenomenologists use the term *lifeworld* to describe this everyday taken-for-grantedness, which, almost always, people in the natural attitude are unaware of reflectively (Finlay 2011; Seamon 2013). One major aim of phenomenology is to study the lifeworld's taken-for-grantedness explicitly. Phenomenologists have become progressively interested in the phenomenon of place because it is a primary contributor to the spatial, environmental, and temporal constitution of any lifeworld, past, present, or future. As phenomenological philosophers Edward Casey (1997, 2009) and Jeff Malpas (1999, 2006) have powerfully demonstrated, *human being is always human-being-in-place.*

In speaking of a lived inseparability between people and the places in which those people find themselves, Casey, Malpas, and other phenomenologists contend that place is not a material or geographical environment distinct from human beings but, rather, the indivisible, normally taken-for-granted phenomenon of person-or-people-experiencing-place.[3] These phenomenologists emphasize the role of the

[2] Introductions to phenomenology and phenomenological methods include: Cerbone 2006; Finlay 2011; Moran 2000; Seamon 2013; Seamon and Gill 2016; van Manen 2014.

[3] Phenomenological discussions of place include: Casey 1997, 2009; Donohoe 2014; Malpas 1999,

lived body and environmental embodiment as integral aspects of place and place experience. This understanding is particularly indebted to Merleau-Ponty, to whose discussion of these topics I now turn.

3 Merleau-Ponty and Perception

In all of his work, Merleau-Ponty asks one central question: How can it be that human beings are present to a world that immediately makes sense?[4] In our ordinary daily experience, we find ourselves in a particular moment of comprehensible experience that necessarily incorporates a taken-for-granted but understandable world in which our experience unfolds. In the sense that this world is there alongside us in a particular way without any necessary effort on our part, we can say we are enmeshed and entwined in our world, which, simultaneously, is enmeshed and entwined in us. Some manner of world is always present, whether that world is as small as a telephone booth or as expansive as the vista from the Grand Canyon's rim. Every moment of our lives, we always find ourselves caught up and immersed in a world that is there before us in inescapable presence.

Merleau-Ponty argues that the experiential foundation of this immersion-in-world is *perception*, which he relates to a *lived body* that simultaneously experiences, acts in, and is aware of the world that, typically, responds with immediate pattern, meaning, and contextual presence.[5] Merleau-Ponty understands the lived body as a latent, lived relationship between an intentional but pre-reflective body and the world it encounters and perceives through continuous immersion, awareness, and actions. In this sense, perception is the *immediate givenness of the world founded in corporeal sensibility* (Cerbone 2008, pp. 128–131). Merleau-Ponty (1962, p. 58) contends that perception is a foundational, always-present quality of human experience and meaning but difficult to grasp and articulate intellectually for two reasons, first, because its presence and significance typically lie beneath conscious cerebral awareness. Second, by its very nature, perception places itself in the background as it draws us out into the happenings of our world: "perception hides itself from itself […] it is of the essence of consciousness to forget its own

2006, 2009; Mugerauer 1994, 2008; Relph 1976, 2009; Seamon 2013, 2014, 2017, 2018; Stefanovic 2000.

[4] Introductions to Merleau-Ponty's philosophy include: Carman 2008; Casey 1997, pp. 228–242; Cerbone 2006, pp. 96–133; Diprose and Reynolds 2008; Moran 2000, pp. 391–434; Pietersma 1997: Romdenh-Romluc 2012. Works relating Merleau-Ponty's thinking to environmental, spatial, and place themes include: Allen 2004; Casey 2009, pp. 317–348; Cataldi and Hamrick 2007; Evans 2008; Hill 1985; Leder 1990; Locke and McCann 2015; Morris 2004, 2008; Seamon 1979, 2013; Toombs 1995, 2000; Weiss 2008; Weiss and Haber 1999.

[5] Discussions of Merleau-Ponty's understanding of the lived body include: Behnke 1997, pp. 66–71; Bermudez et al. 1999; Casey 1997, pp. 202–242; Cerbone 2006, pp. 96–103, 2008; Evans 2008; Finlay 2006; Heinämaa 2012, pp. 222–232; Jacobson 2010, pp. 223–224; Leder 1990; Moran 2000, pp. 412–430; Morris 2004, 2008; Pallasmaa 2005, 2009; Seamon 1979, 2013, 2015.

phenomena thus enabling 'things' to be constituted" (Merleau-Ponty 1962, p. 58). Merleau-Ponty's central aim is to reconsider perception phenomenologically by "reawakening the basic experience of the world" (1962, p. viii).

In *Phenomenology of Perception,* Merleau-Ponty claims that conventional philosophy and psychology have misunderstood perception, introducing erroneous, reductive concepts like sensations, stimuli, judgments, or cognitive representations that misrepresent and therefore distort any conventional philosophical or scientific accounts of actual perceptual experience. In everyday life, the world is not engaged through mental images or separate sensory units somehow translated into more organized meanings. Instead, the world we encounter makes immediate sense as all its parts – whether objects, living beings, situations, or events – fit readily into place. In this way, experience can be pictured as an interpenetrating web of sensory and bodily presence and relationship – what Merleau-Ponty (1962, p. 4 & 15) identifies as the *perceptual field*: "The perceptual 'something' is always in the middle of something else, it always forms part of a 'field'" (Merleau-Ponty 1962, p. 4).

This perceptual field means that our lived awareness is not the sum of isolated sensory inputs or cognitive representations but a dynamic and unique commingling of integrated lived possibilities in each moment of experience. Conventional philosophy and science have regularly sought to define and understand the five senses separately (Carman 2008, pp. 67–74), but Merleau-Ponty emphasizes that, as perceptual field, the senses intermingle and mutually resonate. The result is what he calls "synaesthetic perception" – "a whole already pregnant with an irreducible meaning" (Merleau-Ponty 1962, p. 229, pp. 21–22). He claims that this lived synergy of the perceptual field is grounded in the sensibilities and possibilities of the lived body that, by its very nature, evokes and engages meanings from the world: "My body is the fabric into which all objects are woven, and it is, at least in relation to the perceived world, the general instrument of my 'comprehension'" (Merleau-Ponty 1962, p. 235). In other words, qualities of the world directly resonate with the lived body and thereby convey immediate meanings and ambiences, though typically at a tacit, unself-conscious level of awareness that is best located and described via phenomenological study.

4 Merleau-Ponty and Body-Subject

If perception for Merleau-Ponty involves a passivity of sensory experience, he also identifies a more active *motor* dimension, associated with bodily movement, actions, and skills. In the realm of everyday experience, these sensory and motor facilities of the lived body are never separate but, through the perceptual field, work together seamlessly so that awareness and action unfold as an integrated, continuous experience (Carman 2008, pp. 78–79). Merleau-Ponty understands everyday bodily mobility in terms of a *body schema,* or *body-subject,* as I call it here. Body-subject can be defined as a pre-cognitive, bodily intelligence and intentionality manifested through action and intertwining with the world at hand. As Merleau-Ponty (1962, pp. 138–39) explains, "Consciousness is being towards the thing through the intermediary of the body."[6]

In relation to place experience, a central concern is how body-subject might be understood in relation to larger-scaled corporeal movements unfolding in relation to rooms, buildings, streets, public open spaces, neighborhoods, and the like. In *Phenomenology of Perception,* Merleau-Ponty provides several real-world examples of how body-subject automatically adjusts larger-scale movements so there are no disruptions or accidents: a lady's accommodating a hat with a feather; a blind man's using his walking stick; a motorist's driving his automobile (1962, pp. 143–146). Merleau-Ponty's most significant example is a description of his bodily mastery of the apartment where he lived: "My flat is, for me, not a set of closely associated images. It remains a familiar domain round about me only as long as I still have 'in my hands' or 'in my legs' the main distances and directions involved, and as long as from my body intentional threads run out towards it" (1962, p. 130).

Drawing on Merleau-Ponty, other studies have pointed to the spatial versatility of body-subject as expressed in more complex bodily ensembles extending over time and space and fashioning a wider lived geography (Allen 2004; Hill 1985; Seamon 1979; Toombs 1995, 2000). As philosopher Kirsten Jacobson (2010, p. 223) explains,

> the body finds a way to gear onto the world that allows it to find its way successfully within that world; once developed we are significantly carried through our days by this now-habitual attunement. Thus the body's habits and its developed orientation reflect a certain style of *having a world* and also create a zone of familiarity within which we carry out our daily dealings….

In my own work (Seamon 1979), I have identified two specific bodily ensembles grounding this "zone of familiarity": first, *body-routines* – sets of integrated gestures, behaviors, and actions that sustain a particular task or aim, for example, tying a bow, washing dishes, making a sandwich, doing home repair, and so forth; and, second, *time-space routines* – sets of more or less habitual bodily actions that extend

[6] I prefer "body-subject" to "body schema" because "subject" better suggests than "schema" the pre-reflective but intelligent *awareness* of the lived body. Whichever term one uses, it is important to recognize that the underlying imagery of corporeal autonomy (i.e., body as autonomous agent) can readily lead to another kind of subject-object dichotomy and bifurcation that loses sight of the lived connectedness and intimacy between body and world. As Morris (2004, p. 36) explains:

> Unfortunately, it is all too easy to reify the body schema [body-subject], to conceive it as an independent thing, a bridge built in advance that is to be abstracted from the movement in which it emerges […] Once we have an 'it', a schema, to talk about, our tendency is to turn it into a thing, because our minds and languages – and the body schema itself – disposes us to lend a thingly, solid *sens* [sense] to the content of the world. Merleau-Ponty himself does not escape this tendency and sometimes even invites misconception of the body schema as some sort of thing.

In this chapter, I work to avoid the danger of bifurcation, but it is extremely difficult to keep to an understanding and language that faithfully depict the intertwinement of body and world as a lived whole rather than as a body/world duality. The same is true of the lived entwinement and interconnectedness of people and place.

through a considerable portion of time, for example, a getting-up routine or a weekday going-to-lunch routine. In a supportive physical and spatial environment, individuals' bodily routines can intermingle in time and space, thereby contributing to a larger-scale environmental ensemble that I have called, after the earlier observations of urban critic Jane Jacobs (1961, pp. 50–54), a *place ballet* – an interaction of individual bodily routines rooted in a particular environment, which often becomes an important place of interpersonal and communal exchange, meaning, and attachment, for example, a well-used office lounge, a popular tavern, a lively city street, a robust urban plaza, or a thriving city neighborhood (Fullilove 2004; Oldenburg 2001; Seamon 1979, 2013, 2018; Seamon and Nordin 1980).

5 Interpretation 1: A Passage from García Márquez's *One Hundred Years of Solitude*

To illustrate how Merleau-Ponty's argument is eminently applicable to everyday human experience, I draw on a descriptive passage from García Márquez's *One Hundred Years of Solitude* (1967/1970, pp. 246–248). In this passage, García Márquez depicts the situation of elderly family matriarch Úrsula Iguarán, who is losing her sight because of cataracts. She refuses to reveal her condition to family members, however, because blindness would be a sign of uselessness in her remote village community. She sets herself to cope by focusing on "a silent schooling in the distances of things and people's voices," so that she can see, via astute mental recollection, what her failing eyes cannot. She becomes keenly aware of odors, which intensify her environmental sensibilities in a way much more acute than the visual presence of "bulk and color." She comes to know so well the placement of everything in her everyday world that she forgets much of the time she is blind.

In illustrating Úrsula's re-mastery of her lifeworld García Márquez describes two experiences, the first involving family member Fernanda, who loses her wedding ring. Having carefully noticed, because of her blindness, the daily behavior patterns of other family members, Úrsula easily locates the ring. She has come to recognize a household regularity of which other family members are unaware because that regularity is a taken-for-granted part of the lifeworld and out of sight because of the natural attitude. The second experience illustrating Úrsula's re-mastery of everyday life is an afternoon mishap in which Úrsula collides on the porch with another family member, Amaranta, who is sitting there sewing. Amaranta chides Úrsula for her carelessness, but Úrsula retorts that it is Amaranta's fault because she is not sitting in her usual place. After this incident, Úrsula attends to Amaranta's porch behaviors more carefully and realizes that, because of the sun's shifting seasonal angle, family members imperceptibly changed their porch positions without being aware: "From then on Úrsula had only to remember the date to know exactly where Amaranta was sitting."

In recasting García Márquez's account phenomenologically, we can say that, via the paradox of Úrsula's not being able "to see," she necessarily must make her lifeworld an object of directed attention, though, phenomenologically, it is important to point out that she makes this effort from within the natural attitude, worried about repercussions if other family members discover her blindness. She re-masters her household lifeworld, first, by reconstructing a viable perceptual field sustained by her remaining sensibilities, particularly odor and an intuitive sense of distances and locations. In addition, she astutely attunes herself to family members' everyday movements and actions, thereby shifting her attention to what might be described phenomenologically as the "bodily constitution of everyday place." From Merleau-Ponty's perspective, she progressively becomes aware of the habitual regularity of household body-subjects.

Specifically, this realization for Úrsula centers on her discovery that household members unself-consciously partake in repetitive actions and routines. She understands, for example, that Fernanda's ring can be found through attention to the "only thing different that she had done that day" – airing the mattresses, removing her ring, and putting it in a place away from the children. Similarly, because Amaranta had shifted her porch position to be in the sun, Úrsula realizes how the absolute regularity of household routines work in dialogue with shifting aspects of place, in this case, seasonal change. This example illustrates how a physical constituent of place – the shifting sun – contributes to a habitual, unnoticed shift in how human beings respond to their place. It is a subtle but vivid example of the mutual given-and-take between human beings and the world in which they are immersed.

In describing a house-wide interconnectedness between people and their place, García Márquez's account intimates the spatial versatility of body-subject as expressed in more complex bodily ensembles extending over time and space and fashioning a wider household dynamic, including *body-routines* as illustrated by Úrsula's making breakfast for the family or repairing a torn shirt; and *time-space routines* as illustrated by Úrsula's realization of Amaranta's shifting porch position. As García Márquez summarizes the situation, Úrsula "discovered that every member of the family, without realizing it, repeated the same path every day, the same actions, and almost repeated the same words at the same hour."

As indicated by Úrsula and Amaranda's collision on the porch, García Márquez also suggests that the household incorporates a good amount of *place ballet* in which habitual actions of individual family members commingle to make the household a network of interconnected movements and encounters. Out of individual bodily actions automatically intermingling in space unfolds a more complex synergistic structure that marks Úrsula's home world as a dynamic wholeness facilitating and being facilitated by place. As Casey (2009, p. 327) explains, "lived bodies belong to places and help to constitute them" just as, simultaneously, "places belong to lived bodies and depend on them."

6 Interpretation 2: Observations of a Walk Between Home and Work

This co-constitution of lived bodies and place is powerfully depicted in García Márquez's account of Úrsula's household, which, in archetypal fashion, illustrates Casey's contention that lived bodies and places belong together in mutual relationship and support. I now turn to the observations of my weekday walks to consider other lived aspects of place. I start with Merleau-Ponty's claim that the lived body grounds everyday movement (1962, pp. 137–138) but extend this claim further via Casey's contention that "Part of the power of place, its very dynamism, is found in its encouragement of motion in its midst" (2009, p. 326). In refining Merleau-Ponty's understanding of the lived body and motility, Casey (2009, pp. 326–327) identifies three kinds of bodily movement relating to place: first, the body remaining in place but moving, for example, Úrsula's preparing food on the stove; second, the body moving within a place, for example, her family members' household traversals; and, third, moving between places, for example, my weekday walk that I examine more fully here.

This walk between home and work usually takes 20 min and is about one mile in length along five streets and university sidewalks. The route crosses two well-trafficked city streets and four less-trafficked neighborhood streets. Other than a short stretch through the university campus, the walk traverses three different neighborhoods, two of upper-middle-class housing and one of fraternities and student apartments. There are no shops or eateries on my route. Other than along the stretch through my university campus, I typically meet few other pedestrians as I walk back and forth in the morning, usually around 9 am; and in the evening, usually around 6 pm.

Here, I consider this walk phenomenologically, drawing on seventy-four of the ninety-five observations that I made of this walking experience over the course of several months.[7] First-person observation and description of one's experience is a legitimate phenomenological method, and my aim in recording observations was to assemble a range of lifeworld experiences that I could then examine for broader experiential themes underlying the original descriptions.[8] Because phenomenological method demands an openness to the phenomenon, I followed no organized

[7] Originally, my plan was to use these observations as one set of experiential descriptions for the conference presentation mentioned in footnote 1. I began the observations shortly after I accepted the conference invitation, in early December, 2014. Ultimately, I did not draw on the observations for my conference paper but decided to incorporate them in the present chapter. I therefore continued observations of walks right up until the time of writing this chapter – early September, 2016.

[8] Merleau-Ponty (1962, p. viii) points toward the value of first-person explication when he writes that "We shall find in ourselves, and nowhere else, the unity and true meaning of phenomenology. It is less a question of counting up quotations than of determining and expressing in concrete form this *phenomenology for ourselves*" On the use of first-person accounts in phenomenological research, see Finlay 2011; van Manen 2014. One of the best examples of first-person explication is the work of philosopher Kay Toombs (1995, 2000), who lives with multiple sclerosis and explores her illness phenomenologically via concepts like the lived body.

method for gathering observations. Instead, as I walked, I would suddenly become aware of some event or situation toward which I then worked to become more alert and present. If I remembered when reaching office or home, I would go to the computer and type my experience as I remembered it. Often, I would note an event but forget to record it. And during most walks, I would lose sight of the task entirely and make no observations, thus no doubt missing situations and experiences that otherwise might offer phenomenological insight.

In this sense, the observations I draw on here are serendipitous, and a set of descriptions gathered at other times, past or future, would probably be considerably different as specific experiences. From a phenomenological perspective, however, one can argue that my ninety-five observations are one representative portrait of my weekday walk and an acceptable starting point for locating broader phenomenological themes relating to place and movement. Here, I discuss three interpretive themes, all of which relate in some way to the lived body, including an emotional dimension. These three themes are: (1) body-subject and perceptual field; (2) intercorporeal presence; and (3) modes of place encounter. Of the seventy-four observations that provided the descriptive field out of which these three themes arose, I draw only on forty-six that are most illustrative of these three themes. Mostly in the discussion below, I summarize these observations; the appendix provides the complete text of these forty-six observations, numbered in the order they are discussed.[9]

1. **Body–Subject and Perceptual Field**

Twelve of the forty-six observations in the appendix relate to some aspect of body-subject, particularly situations in which habitual movements are interfered with in some way. For example, I enter the east wing of my university building, forgetting that I have just moved to a new office more conveniently accessed via the building's west wing (observation **1**). My usual campus route is upset by a construction detour, and I feel a moment of irritation because I am required to go "out of my way" (**2**). On my way home after a four-inch mid-afternoon snowfall, I walk in the plowed streets rather than on the unshovelled sidewalks. I encounter two cars stopped and blocking my passage along the street. I am annoyed because, to get around the cars, I must move to the snow-covered sidewalk that is much more

[9] The three interpretive themes that I identify and discuss here arise out of careful, multiple readings of the original ninety-five observations recorded during the several-month period in which I set myself to pay heed to my weekday walks. Twenty-one observations do not relate to the three interpretive themes and are thus not discussed. Of the seventy-four observations that are the descriptive base out of which I located these three themes, twenty relate to "body-subject and perceptual field"; twenty-eight to "intercorporeal presence;" and twenty-six to "place encounter." As already explained, I draw on only forty-six of these seventy-four observations because I judge these forty-six observations to provide sufficient real-world evidence for claiming the validity of the three interpretive themes, which owe much to Merleau-Ponty's broader phenomenological discoveries and principles. Depending on their conceptual perspectives and topical interests, other phenomenologists might generate a different set of observations and locate different interpretive themes. The key phenomenological concern as to relative interpretive trustworthiness is the degree of correspondence and fit between descriptive accounts and interpretive generalizations; see Seamon 2017; van Manen 2014; Wachterhauser 1996.

difficult to traverse (**3**). Some observations detail how quickly and efficiently cognitive awareness works out a new route when body-subject's habitual route is interfered with. I'm heading home and see that, during the day, construction workers have erected a chain-link fence across the sidewalk I normally use. I quickly decide to use another sidewalk I traverse only when I drive to school and walk to and from the parking lot (**4**). My usual street is blocked by an accident and the police won't allow entry: "I note annoyance and quickly think out a way to shift my route to get home" (**5**).

Other observations point to the continuous, integrated interconnectedness between body-subject and perceptual field. I realize my shoelace is untied and scan for a secure raised surface on which I can rest my foot comfortably as I retie my shoe (**6**). A cold fall wind blows on a sunny October day, and I find myself walking on the warmer, sunny portions of sidewalk (**7**) It's a hot summer day and I find myself seeking out stretches of shaded sidewalk (**8**). My way is blocked by new construction fencing but I don't wish to retrace my route. As I look for a way to move around the fencing, I find myself "automatically pointing my body toward a space in the hedgerow separating the parking lot from the lawns" (**9**). In some observations, the perceptual field includes a technological component, which in my case is sometimes listening to an ipod on which I have loaded favorite songs. For example, I'm walking home in the dark, listening to Peter Gabriel's "Solsbury Hill." Its rhythm synchronizes perfectly with my footsteps, and I "do a kind of dance as I walk, swinging arms from side to side" (**10**). I'm returning home and listening to Don Henley's "The End of the Innocence." I can't find the right walking rhythm but then realize the trick is shorter steps: "Such a pleasure to 'walk dance' in rhythm to a fine song!" (**11**).

The most comprehensive observation relating to the unbroken, effortless intermesh between body-subject, movement, and perceptual field involves my being caught in a sudden spring rainstorm (**12**):

> The sidewalks and streets are suddenly awash in water. I am struck by how my eyes pay attention to the water puddles that my feet jump over as my hand adroitly repositions my umbrella at the angle most effective for deflecting the pelting rain. I feel wetness as water splatters my ankles and hear and smell the rain as it strikes the earth. I observe how my attention continuously shifts in an automatic way, immediately aware of the next water puddle I must move around or the sudden awareness that my left shoe is soaked because I gauge the flow of water along the street as shallower than it actually is. There is a moment when I notice daffodils blooming along a picket fence I always walk by, and another moment when I notice a pedestrian running up the street for the cover of his parked car. So much happening in such an ordinary event!

As this observation suggests, the experience of walking through the rain involves the perceptual field's continuously shifting fabric of sensory and bodily awareness and action that *just happen*. At each moment, some parts of this field are more prominently present in experience – for example, the next water puddle I must traverse or the sudden awareness that my left shoe is soaked. But at each instant, the more focused portion of my experience remains inescapably conjoined with the momentary, less central, aspects of the situation – the daffodils blooming along the picket fence or the pedestrian running up the street for the cover of his parked car. In short, perceptual experience always involves a continuously shifting figure and ground brokered by a broader constellation of related significances and actions all coordinated instantaneously with body-subject. This lived integration of movements, surroundings, and need to get home or to work is described most broadly by Merleau-Ponty as a dynamic bodily process via which human beings are always already situated:

> [M]y body appears to me as an attitude directed towards a certain existing or possible task [...] The word "here" applied to my body does not refer to a determinate position in relation to other positions or to external co-ordinates, but to the laying down of the first co-ordinates, the anchoring of the active body in an object, the situation of the body in face of its tasks. Bodily space can be distinguished from external space and envelop its parts instead of spreading them out because it is the darkness needed in the theatre to show up the performance. (Merleau-Ponty 1962, pp. 100–101).

2. Intercorporeal Presence

Seventeen of the forty-six observations in the appendix describe interpersonal encounters that range from momentary bodily co-presence to lengthy conversational exchange. Some observations depict an unself-conscious effort to maintain a comfortable distance between myself and nearby others who are moving or standing. Exiting a building, a woman walks right in front of me and "To make some distance between us, I walk perpendicular to her direction of movement and then meander a bit to make more space" (**13**). I note a group of five college students crossing the street ahead of me, and "I take my time so there is some space between them and me" (**14**). A group of school children turn the corner onto the street I'm walking, and I slow my pace "so I will have some space. I don't fancy being right behind noisy, unruly kids." At the next stoplight, I wait a bit so there is "even more space between them and me" (**15**).

Some observations illustrate how firsthand encounter with others can precipitate discomfort or other negative feelings. I walk on the sidewalk and a man with an unkempt appearance approaches from the opposite direction. I note a "mild feeling of discomfort with the encounter to come. We pass, acknowledge each other with a hello, and continue on our way" (**16**). I'm walking down the hill from my house and, suddenly, I sense a cyclist speeding by, very close. I am annoyed and angered: "What if I had suddenly veered left to avoid something on the pavement? It's a steep hill. He's going too fast!" (**17**).[10]

The largest number of observations involving intercorporeal encounter relate to moments when my walk is interrupted or interfered with by others co-present in some way. I'm walk home on graduation day, and there are many street parties on my route. I note ahead two young men blocking "my" sidewalk. "An inner voice

[10] The state of Kansas is stereotypically imagined as flat topographically, and this is the case for western Kansas. My home city of Manhattan, however, is part of a natural region known as the "Flint Hills," a tall-grass prairie with rolling hills, of which some are quite steep, including the hill on which I live.

says, 'You're in my space, get out of the way'." They sense my approaching and sidle aside to the adjacent lawn. I note that they don't look at me directly but must sense my movement in their peripheral vision (**18**). A driver doesn't signal at an intersection and turns into the street I've just begun to cross, figuring he is continuing straight ahead (**19**). I cross a busy street and have the light, but a truck comes from the opposite direction and cuts in front of me to beat the turn light. I shout, "damn driver!" and am visibly angry (**20**). I'm walking on the sidewalk in front of an apartment-complex entry, and a young woman in a SUV turns into the entrance without signal, cutting in front of me. There is a large pothole filled with rainwater, which splashes up toward me, but I quickly back step and don't get wet. I note a cascade of feeling: irritation that she didn't signal, worry that I may get wet, anger that she did not have the courtesy to let me pass before she turned into the entry (**21**).

There are other observations that illustrate more supportive modes of interpersonal encounter. I pass two lovers saying goodbye at their car door, and I "turn my head and attention away, projecting a sense of "not being there" so as not to interfere with the privacy of their moment together" (**22**). I encounter coming toward me an older couple walking their dogs, and I move into the street so my passing won't be too close to upset the canines (**23**). I cross a busy street and wonder if the school bus driver, turning right, will allow me right of way. He does, and I wave my hand in thanks (**24**). I pass a young father and his son and say hello. They both say hello in return. I think that "it's good that the dad acknowledge a passerby and set a good example for his son" (**25**). In contrast, I pass a young father and his son but am rebuffed when I say hello. I write that it "seems a poor way to introduce a child to his neighbors and the public realm" (**26**).

Three observations involve conversing with others I meet on my walk. An acquaintance is doing yardwork as I pass, and we discuss the new plants and shrubs he's using as ground cover for a replaced sewage line (**27**). A scruffy, bearded man waves his arms at me across the street, but I can't hear what he's saying. At first I think he's a homeless panhandler, but, as I get closer, I realize he's asking me the location of the nearest Burger King eatery. I give him directions and he tells me he's visiting his mother-in-law and doesn't know the city (**28**). I pass the house where I used to greet an old gray cat every day. I notice that his owner is out trimming weeds and ask her about the him, since I haven't seen him for several days. She tells me that he died on July 4th and she buried him in the back yard. I relate to her how he had been an important "event" in my walks and thanked her for taking him in. He had had a hard life before she took on his care, and I feel sad as I walk on (**29**).

In *Phenomenology of Perception,* Merleau-Ponty devotes a chapter to "Other Selves and the Human World," which considers how people are co-present in the world. Just as we are immersed in worlds via perception and body-subject, so we are immersed in worlds that include other human beings. Merleau-Ponty writes that we experience the social world, "not as an object or sum of objects, but as a permanent field or dimension of existence [...] Our relation to the social is, like our relation to the world, deeper than any express perception or any judgment [...] The social is already there when we know it or judge it" (Merleau-Ponty 1962, p. 362).

We see this preset co-presence in the observations above: The lovers by the car; the dog walkers; the father and son acknowledging me on the sidewalk; the bus driver allowing me the right of way; the man asking directions. These "others" are already present in my lifeworld, and I reciprocate with a spontaneous, matter-of-fact response: turning away; moving aside; saying hello; waving a hand of thanks; giving directions. As Merleau-Ponty writes, "Round about the perceived body, a vortex forms, towards which my world is drawn and, so to speak, sucked in […] Already the other body has ceased to become a mere fragment of the world, and become the theatre of a certain process of elaboration" (Merleau-Ponty 1962, p. 353). Like the habitual actions of my own body-subject, these others are an integral constituent of my lifeworld's dynamic, intertwined fabric. Phenomenologist Elizabeth Behnke (1997, p. 69) calls this situation *intercorporeality*, which she describes as "a primordial solidarity of and between embodied beings – a pre-personal communality that is never fully effaced but sustains a reciprocal interplay of one's own and others' comportment prior to any explicit consensus."

As the above observations illustrate, this immediate co-presence of others often incorporates an emotional cast, which Merleau-Ponty identifies as another dimension of the lived body directing and sustaining our lived relations with the world at hand. This emotional presence "can only be grasped through the body and [is] communicated through a reciprocity of intentions and gestures discernable in conduct" (Cataldi 2008, p. 171). On one hand, I feel a tinge of community pride as I exchange hello with the father and son who acknowledge my passing co-presence. On the other hand, I feel a tinge of community despair as the other father and son ignore me. I encounter the scruffy, bearded man waving his hands and my feelings shift from discomfort to uncertainty to relief. In short, these lifeworld emotions are one stratum of a "living embodied meaning […] We live [feelings] (rather than know) them. They are a way of establishing embodied relations with the world" (Cataldi 2008, p. 164 & 169).

3. Modes of Place Encounter

Seventeen of the forty-six observations in the appendix relate to what I call *encounter*, which I define as "any situation of attentive contact between the person and the world at hand" (Seamon 1979, p. 99). Unlike the pre-conscious sensibility associated with perception and body-subject, encounter involves some degree of *conscious* awareness – in this case, directed attention to some material, environmental, or human aspect of place as registered during my walks. Observations point toward encounter's range of attentive intensity – from momentary awareness to a deeper, more lasting attention.

Some encounters are invoked by puzzlement: from a distance, I see a gray-brown form on the street ahead that looks like a dead bird, but I come closer and realize the "bird" is a "twisted piece of tree branch wrapped in dead leaves" (**30**). I notice what appears to be a blanket strung between two trees but then realize that the "blanket" is a hammock in which a young man is asleep (**31**). I note a blue notebook left on a retaining wall and wonder why it is so much "out of place" (**32**). I walk by a hillside usually covered with wild flowers and am surprised that it has been cultivated and reseeded (**33**). I encounter a family walking on the sidewalk at the odd hour of 9:30 a.m. and wonder why they're here – they seem out of place on a Friday morning. I watch them enter the nearby Catholic Church and realize they are attending a "Good Friday" service (**34**). Another set of observations illustrate encounters relating to a sense of community responsibility: I notice a mailbox door that has sprung open and close it (**35**); I see a large rusted bolt in the middle of the street and move it to the curb (**36**); I note a branch blocking part of the sidewalk and carry it to the adjacent lawn (**37**).

As with intercorporeal presence, some observations on encounter incorporate an emotional stratum. I notice a dead squirrel at the bottom of my hill, and feel a combination of distress and irritation that drivers can't make more an effort to slow down (**38**). I walk by a dilapidated house in front of which a dispirited puppy has been chained the last few days. I remember to check on him because I am concerned about his wellbeing, but I see he is not there. I feel concern and wonder where the puppy is (**39**). I pass a beautifully crafted, miniature snow man standing on a university stone wall and feel joy that someone would take such care in making such a transient but elegant object (**40**). I walk down my hill on an unseasonably warm February day and am surprised to see, in the open space with trees, a person sitting on a blanket and reading. No one ever uses this space and "I wonder who this person is. Rather mysterious!" (**41**).

Some observations demonstrate how an encounter only happens because something in the world actively triggers my awareness that would not be engaged otherwise. I'm walking to school and a cyclist calls out my name. I look to see who the person is, but I can't tell because he wears sunglasses. I ask, "Who is it?" and the rider answers "Jeff," a neighbor who lives close by (**42**). As I approach home, I am surprised by how barren and less attractive the front yard of my house looks because of the loss a day before of an old maple tree toppled by high winds (**43**). I'm listening to Eliza Gilkyson's "Requiem," and its beautiful melody and haunting lyrics trigger my looking at the flower-covered hillside, the loveliness of which merges with the poignancy of the song (**44**). Campus construction requires that I find a new route to my office and I walk along a stretch of sidewalk I've not traversed before. Its special atmosphere pulls me out of myself and I attend more closely to the experience: "It is the trees that make this place special. There are many, and one moves through a tree-canopied enclosure. The sidewalk meanders. The overall feeling is pleasant, comfortable, reassuring – quite wonderful!" (**45**).[11]

[11] Elsewhere, I have identified this mode of attention to the world as *noticing* – an encounter in which "a thing from which we were insulated a moment before flashes to our attention" (Seamon 1979, p. 108). I identified two modes of noticing: on one hand, "world-grounded noticing," where something in the world "grabs" one's attention (e.g., the miniature snowman); on the other hand, person-grounded noticing, which involves the person's being engaged with things that interest the person – e.g., the birdwatcher's "knack" for noticing birds (p. 108). The observations from my walks illustrate some other ways of depicting noticing phenomenologically: on one hand, becoming aware of something in one's world by chance (e.g., observations 30–34 above); on the other hand, becoming aware of something in one's world because that "something" actively brings attention to itself and thus one become aware of it (e.g., observations 42–45 above). The modes of

One observation is unique in that it describes an encounter in which I sense how all the many independent parts of a place coalesce into a larger, interconnected whole. I note how several individuals all converge in the same moment and space to enable a momentary togetherness that speaks to a sense of place:

> I walk to school and approach the corner of Denison and Leavenworth Streets. I'm walking down the hill of Leavenworth and note a boy in an orange shirt resting against the anchoring post of the stonewall at the corner. A young woman with a tiny terrier walks up Denison on the other side of that street. A blue truck stops at the corner, and the boy runs across the street to the truck. He must have been waiting for a ride. A car coming down Denison slows for the stopped truck. The boy gets in and the vehicle speeds away. I see irritation on the face of the driver, who obviously is not pleased that she's had to wait for the truck. The girl with the dog crosses Leavenworth and heads north on Denison. Odd how all these folks converge at that one corner and then disperse (**46**).

From Merleau-Ponty's perspective, these observations on encounter illustrate another way experientially in which the lived body is immersed in place. As we engage with place, we do not, first of all, see and know that place as a detached object via cognitive consciousness. Instead, these things, people, and situations are already present via perception, and we respond in a manner that phenomenologist David Cerbone (2008, p. 129) describes as "getting a grip" on the world we encounter:

> I use the word "grip" here both literally and figuratively as when I grip the pen, coffee cup, hammer, steering wheel [...] in my hands (literally), and when I "get a grip" on things and situations, putting things in order, getting things under control,, and optimizing my perceptual access (figuratively).

Though they are far from a comprehensive picture of how human beings encounter their worlds, the observations I present here do detail some of the ways experientially in which we engage the world with attention and thereby "get a grip" on what that world offers via its already-present perceptual givenness. The moments in which the world comes to presence may be related to puzzlement (wondering about the blue notebook on the stone wall), inconsonance (noticing the open mailbox door and closing it), or unusualness (appreciating the miniature snowman on the stone wall). The observations indicate a range of intensity in how we encounter the world: from momentary awareness (the birdlike branch and leaves) to more intense involvement that often incorporates an emotional dimension (walking along the tree-canopied sidewalk or learning the old gray cat is dead).

It is important to point out that, in most of my walks, encounter was sporadic as body-subject navigated the walk and my attention remained mostly focused inwardly on plans for the day, worries, or random thoughts. I can offer no observations of this obliviousness to the world at hand, since by its very constitution, it assumes a lack of awareness and is therefore largely immune to the attentive contact that my observational method presupposes. But even if, in our typical lives, we notice our world only fleetingly, we still readily cope because of body-subject's

human encounter are complex and multifaceted. What is needed is a thorough phenomenology that includes but moves beyond Merleau-Ponty's unself-conscious perception and body-subject.

pre-reflective awareness. As Merleau-Ponty (1962, p. 146 & 104) summarizes the situation, "The body is our general medium for having a world [...] [It is] the matrix of habitual action [and] a means of ingress into a familiar surrounding."[12]

7 Place, Environment, and Situatedness

Phenomenologist Dermot Moran (2000, p. 391) perspicaciously pinpoints the crux of Merleau-Ponty's thinking as "the 'mysterious', 'paradoxical', 'ambiguous' nature of our embodiment in a world that seems pre-ordained to meet and fulfill our meaning-intended acts." Merleau-Ponty's aim is "the reawakening of an understanding of the original acts whereby humans come to awareness in the world" (Moran 2000, p. 401). His way of understanding corporeal situatedness is difficult to grasp because we have been educated in a dualistic Cartesian tradition assuming that the world shapes human beings, or that human beings shape the world, or that the two mutually shape each other. In all his work, Merleau-Ponty attempts to circumvent any dualistic conception, since always because of the lived body, people and world are soldered together. In Western philosophy, they have been envisioned conceptually as two but, existentially and experientially, they are always one and undivided.

How might we better clarify the value of Merleau-Ponty's thinking in understanding place? One helpful explication is philosopher Shaun Gallagher's interpretation (Gallagher 1986) of how Merleau-Ponty understands "environment," which, in the argument I have presented here, could also be understood as "place." Gallagher points out that, on one hand, the lived body incorporates the environment: "Phenomenally, the environment is precisely a 'manipulatory area' for the lived body – something potentially to be taken up and incorporated" (Gallagher 1986, p. 163). One example of this "incorporation" is Ursula's household, where body routines, time-space routines, and place ballets contribute to the lived body's inhabiting place. In this sense, environments and places are a distance-gathering protraction of perception and body-subject:

> All of these incorporations or embodiments require the same ability to appropriate "boundaries and directions" in a given environment, to establish "lines of force," in short, to organize the environment, "to build into the geographical setting a behavioural one" [Merleau-Ponty 1962, p. 112]. Geographically, or objectively, the environment is distinguished as standing over and against the living body. Phenomenally, or experientially, the environment is an indefinite extension of the lived body (Gallagher 1986, p. 163).

[12] Elsewhere, I have conceptualized encounter as a continuum of awareness extending from "person-environment mergence" (e.g., an intense engagement with the natural world) to "person-environment separateness" (e.g., obliviousness, whereby "the experiencer's conscious attention is not in touch with the world outside but directed inwardly" [Seamon 1979, p. 104]). No matter how oblivious one is to the world at hand, however, the preconscious perceptual facility of body-subject is present, allowing for necessary habitual actions and routines (see Seamon 1979, pp. 97–128).

Gallagher (1986, p.163) also points out, however, that the environment "conditions the body in such a way that the body is the expression or reflection of the environment." In this sense, environments and places appropriate the lived body and thereby contribute to the particular manner through which individuals and groups inhabit their worlds. As Merleau-Ponty writes, the environment is "a living connection comparable, or rather identical, with that existing between the parts of my body itself" (1962, p. 205).

As I have noted, Gallagher speaks of "environment" rather than "place" throughout his discussion. Though I think his interpretation is perceptive and accurate, I have suggested in this chapter that "place" is a more appropriate concept than environment because the latter suggests a separation, objectively and subjectively, between human beings and their worlds. Though Merleau-Ponty never used it directly, place is a concept more apt phenomenologically because it offers a conceptual and applied means for highlighting the lived fact that human beings are always immersed in their world and that one central facet of this immersion is "being emplaced" and situated via place and the lived body. As Casey (2009, p. 327) explains:

> [T]he body is the specific medium for experiencing a [place]. The lived body is the material condition of possibility for the [place] while being itself a member of that same world. It is basic to place and part of place. Just as there are no places without the bodies that sustain and vivify them, so there are no lived bodies without the places they inhabit and traverse […] Bodies and places are connatural terms. They interanimate each other.

Drawing primarily on Merleau-Ponty's understanding in *Phenomenology of Perception*, I have attempted in this chapter to probe this interanimation between people and place, grounding my interpretation in Merleau-Ponty's philosophical insights as they are illustrated in García Márquez's narrative account and in my observations of walks between home and work. I have suggested that places and lived bodies are mutually supportive and that their interpenetration can be more thoroughly understood through Merleau-Ponty's explication of perception and body-subject. At every moment, perception grounds the taken-for-granted presence of the world at hand, just as body-subject assures that actions and ways of being synchronize with the perceptual field, allowing lifeworlds to unfold with little or no mischance or untowardness. As Casey argues, this synchronicity-in-world is further assured via place, which provides a taken-for-granted sphere of orientation, familiarity, and habitual involvement.

Merleau-Ponty's work is almost impossible to master intellectually because the central phenomenon – the tacit, pre-conscious perception of the lived body – is almost always out of sight of conscious awareness. In everyday experience, this latent realm of presence can only be caught in glimpses as when, for example, I suddenly realize how important the fallen maple was to the everyday aesthetics of my house and yard, or I suddenly notice myself walking to the east wing of my university building, even though my new office is in the west wing. For Merleau-Ponty, phenomenology offers a way to bring the latent, undisclosed dimensions of human experience and meaning to direct attention. His thinking reveals the remarkable but pre-predicative sensibilities, understandings, and actions of the lived body as it handily engages the world at hand. Though he says little about the significance of place directly, his perspective does much to clarify its integral relationship with the lived body and human situatedness.

Appendix

Observations of a Walk Between Home and Work

These descriptions are presented in the order they are discussed in the accompanying text.

Observations Relating to Body-Subject and Perceptual Field

1. To get to my office, I approach my university building's east wing, forgetting that my office has been moved to the building's west wing and that I should be accessing entry from the building's west entry (Tuesday, June 2, 2015).

2. I made my usual right turn toward the alumni center, but the sidewalk is blocked because of construction. A sign reads, "Detour – pass through the alumni center." I become annoyed because I must go out of my way. So much campus construction right now! (Monday, August 24, 2015).

3. It has snowed several inches this afternoon, and the sidewalks are high with snow so I walk in the streets, which have been cleared. I notice ahead that two cars are stopped in the street. I'm annoyed because I must move from the street back into the sidewalk, choked with snow. Why are these drivers stopping their cars in the street? I realize that one party is dropping someone off, and the other party is parallel parking but not doing it very well. Perhaps the poor visibility as the snow continues to fall? (Monday, February 2, 2015).

4. I'm headed home and discover, during the day, that construction workers have erected a chain-link fence around the east portion of the old stadium right up to the street curb on 17th Street. This fence blocks the sidewalk I usually walk home along, and I note myself quite annoyed that I must work out a new route. I quickly decide to use the sidewalk I traverse to get to the old-stadium parking lot when I drive my car to work. There is a good amount of irritation that I must change my usual route (Wednesday, February 11, 2015).

5. There is some sort of accident on Denison tonight, and the police won't let anyone pass. I note annoyance and quickly thinking out an alternative route to get myself home. Rapidly, I picture turning left and walking down 17th Street to Poyntz Avenue, then up Poyntz to my street (Tuesday, February 3, 2015).

6. I note my shoe is untied. I look for a "platform" on which to rest my foot while I re-tie the shoelace. There is nothing immediately available. I see ahead the stair leading to the natatorium. I walk to the stair and retie my shoe (Friday, March 26, 2015).

7. It's a chilly October day and, as I walk to school, I find myself using the stretches of sidewalk that are in the sun. I should have worn a jacket! (Thursday, October, 15, 2015).

8. Today is extremely hot and humid. The sun is fierce, and I do everything I can to stay in the shade. Thank goodness there are trees along much of my route and they shade the sidewalk (Wednesday, August 19, 2015).

9. More construction fencing! I've come too far to turn around, so I look for an alternative way through. I find myself automatically pointing my body toward a space in the hedgerow separating parking lot from lawn. I expect we're not supposed to walk through this space, but I am able to get through and continue on my way (Monday, July 20, 2015).

10. It's late and I'm walking home in the dark. I have my ipod on and am listening to Peter Gabriel's "Solsbury Hill," which syncs so well with my footsteps. I do a kind of dance as I walk, swinging arms from side to side. What the heck? It's dark and no one can see! (Monday, March 26, 2016).

11. Listening on the ipod to Don Henley's "The End of the Innocence." I'm having trouble finding a walking rhythm. Ah, yes, I have it: the song requires short steps and a slight lift in the knee. Such a pleasure to "walk dance" in rhythm with a fine song! (Tuesday, May 12, 2015).

12. As I walk home tonight, I am caught in a sudden spring rainstorm. The sidewalks and streets are suddenly awash in water. I am struck by how my eyes pay attention to the water puddles that my feet jump over as my hand adroitly repositions my umbrella at the angle most effective for deflecting the pelting rain. I feel wetness as water splatters my ankles and hear and smell the rain as it strikes the earth. I observe how my attention continuously shifts in an automatic way, immediately aware of the next water puddle I must move around or the sudden awareness that my left shoe is soaked because I gauge the flow of water along the street as shallower than it actually is. There is a moment when I notice daffodils blooming along a picket fence I always walk by. I notice a pedestrian running up the street for the cover of his parked car. So much happening in such an ordinary event! There is a moment when I notice daffodils blooming along a picket fence I always walk by, and another moment when I notice a pedestrian running up the street for the cover of his parked car. So much happening in such an ordinary event! (April 16, 2015).

Observations Relating to Intercorporeal Presence

13. A young woman exits the natatorium and walks directly in front of me along the sidewalk. To make some distance between us, I walk perpendicular to her direction of movement and then meander a bit to make more space. I note how I feel uncomfortable when a person I don't know is too close (Monday, March 21, 2016).

14. There's a group of five male college students crossing the street in front of me, and I take my time so there is some space between them and me (Thursday, October 15, 2016).

15. I'm walking to school and see a class of school children turn the corner onto Denison. I slow my pace so I will have some space. I don't fancy being right behind noisy, unruly kids. At the stoplight at Denison and Anderson, I wait a bit so there is even more space between them and me (Monday, July 20, 2016).

16. There is an unkempt man coming toward me on the sidewalk. I note a mild feeling of discomfort with the encounter to come. We pass, acknowledge each other with a hello, and continue on our way (Tuesday, March 29, 2016).

17. I'm walking down the hill from my house and a cyclist speeds by, almost brushing me. Annoyance and anger – that someone would be so thoughtless as to ride so close to a pedestrian. What if I had suddenly veered left to avoid something on the pavement? It's a steep hill. He's going too fast! (Friday, April 1, 2016).

18. It's graduation day, and there are some street parties in the houses I walk by. There are two young men standing in "my" sidewalk. An inner voice says, "You're in my space, get out of the way." They sense my approach and sidle aside to the adjacent lawn. I note they do not look at me directly but sense my movement in their peripheral vision (Saturday, May 16, 2015).

19. A driver doesn't have his signal on at the intersection of Delaware and Poyntz, so I assume he's continuing straight ahead. I start to cross the street, but he turns in front of me. I stop and wait for him to pass. What a jerk! (Monday, August 21, 2016).

20. I'm crossing Anderson with the walk light for pedestrians. A truck speeds to the intersection and turns left, cutting in front of me even though I have the right of way. I swear and say, "Damn drivers!" (Friday, February 13, 2015).

21. I'm walking to school and, from the other direction, a young woman driving a large SUV turns into her apartment drive that crosses my sidewalk. I've already noticed that she hasn't signaled, and now she cuts in front of me, striking a large pothole full of water that splashes toward me. There is a quick emotional flash: "I hope I don't get wet!" I'm angry because she didn't signal, she cut in front of me, and almost covered me with water. I swear and ponder how stupid people can be (Monday, August 22, 2016).

22. As I walk home, I see lovers saying goodbye at their car door. I walk by the car and turn my head and attention away, projecting a sense of "not being there" so as not to interfere with the privacy of their moment together (Tuesday, May 26, 2015).

23. Walking home on Denison, I see a couple with their two dogs approaching me on the sidewalk. I move into the street just in case my presence might upset the canines. No need to excite excitable dogs! (Monday, March 28, 2016).

24. I'm crossing Anderson toward the university and a big yellow school bus is in the right turning lane. I enter the crosswalk and wonder if the driver will stay stopped or attempt to move right before I get to that side of the street. I note he remains stopped, and I wave my hand in thanks for his giving me the right of way (Friday, February 6, 2015).

25. I'm headed home and approach a young father and son out for a walk. I say hello, and they both say hello in turn. I think it's good that the dad acknowledges a passerby and sets a good example for his son (Wednesday, May 27, 2015).

26. I'm walking to work, and a young father and son move toward me on Denison. I say hello but receive no acknowledgement. The father scowls. Seems a poor way to introduce a child to his neighbors and the public realm (Thursday, March 19, 2015).

27. An acquaintance is doing yardwork as I pass, and we discuss the new plants and shrubs he's using as ground cover for a replaced sewage line. Some of these plantings I need to try in my front lawn. We converse for about 10 min, but I must get home to make dinner (Wednesday, July 13, 2016).

28. I'm walking down my hill and approaching Poyntz. Across the street, a scruffy, bearded man waves his arms at me, but I can't hear what he's saying. At first I think he's a homeless panhandler, but, as I get closer, I realize he's asking me the location of the nearest Burger King eatery. I give him directions and he tells me he's visiting his mother-in-law and doesn't know the city. I feel relieved that I can help him so easily. I was not in the mood to deal with a homeless person this morning (Wednesday, May 13, 2015).

29. I notice that the woman who lives in the house where the old gray cat used to live is trimming weeds. I call out to ask her what happened to the cat, which I would say hello to each day as I walked by. He was very old and had a hard life until this kind-hearted lady took him in. She explains that he died on July 4th and that she buried him in the back yard. I relate to her how he was an important "event" when I would walk by. He conveyed so much suffering and hurt. I felt sad as I continued on (Monday, July 29, 2016).

Observations Relating to Place Encounter

30. I'm walking down the hill from my house. In the distance, I see near the right curb a gray-brown, baglike form that looks like a bird. I come closer and realize the "bird" is a twisted piece of tree branch wrapped in dead leaves. I feel relieved. Not another "road kill." I continue walking (Friday, December 12, 2014).

31. I'm walking home on Denison and suddenly see in my peripheral vision something hanging between two trees that looks like a large blanket. I look more carefully and realize that a man has put up a hammock in which he is resting. He looks asleep (Thursday, May 12, 2015).

32. I'm walking along a stone wall in front of a house on Denison. I spy on the wall a blue notebook that seems so out of place, especially since it looks like it will be raining soon. I think about moving the notebook to the front porch of the house but don't. Who knows who it belongs to?! (Monday, December 15, 2014).

33. I walk by a hillside that is usually covered with wildflowers this time of year, but this spring the owner has cultivated and reseeded the hillside, which is now mostly bare earth. I miss the flowers. They were a pleasure to look at (Wednesday, April 20, 2016).

34. I'm walking down Denison near Poyntz and see, on the opposite side of the street, two adults and three kids – a family. It's about 9:30 am and this group seems out of place. They cross Poyntz, and I realize they are going into the Catholic Church for a service – it's "Good Friday." They must have parked their car on one of the side streets and are walking the rest of the way to the church (Friday, March 25, 2016).

35. A mailbox of a house on North Delaware is open. I close it. I hope someone else would do the same for me (Monday, March 28, 2016).

36. There is a large rusted bolt in the middle of Poyntz. Cars might strike it so I pick it up and move it to the far curb (Wednesday, January 14, 2015).

37. There has been a strong wind this morning. On the Denison sidewalk, I encounter a fallen tree branch blocking the right of way. I move the branch to the adjacent lawn. It would be an obstacle for children walking this way (Tuesday, March 24, 2015).

38. As I walk to work, I see a dead squirrel at the bottom of the hill. Poor creature. Why can't drivers slow down? (Monday, March 16, 2016).

39. I walk down Leavenworth and past a dilapidated house in front of which an unhappy puppy has been chained the last few days. I remember to check on him because I am concerned about his wellbeing, but I see he is not there. I feel concern because he looked so dispirited and uncared for. I wonder what's happened to him? (Friday, March 25, 2016).

40. As I enter campus, I see, on the stone wall next to the sidewalk, a handsomely crafted, miniature snow man with bottle-cap eyes. He really is beautifully made, and I feel a sense of joy that someone would take so much care in making such a transient object (Tuesday, January 19, 2016).

41. I walk down my hill on an unseasonably warm February day and am surprised to see, in the open space with trees, a person sitting on a blanket with a bike propped up nearby. It is a lovely afternoon – at least 60 degrees Fahrenheit – so it makes sense one would wish to be reading outside in the sun under a tree. But I've never before seen anyone use the space this way. I wonder who this person is? Rather mysterious! (Friday, February 6, 2015).

42. I'm walking to school. Someone calls my name, but I don't recognize him because he's on a bike and wears sunglasses. I ask, "Who is it?" and he answers "Jeff." I realize it's my neighbor, but he seems "out of place" with the bike and sunglasses (Monday, July 6, 2015).

43. I return home and note how different my house and front yard look. Yesterday there was a fierce thunderstorm, and the old maple tree near the entry of my house was blown over. The yard looks barren without the tree, and the aesthetic sense of my property is not the same. I'll need to plant a new tree! (Thursday, September 10, 2015).

44. I'm listening on my ipod to Eliza Gilkyson's "Requiem." Its beautiful melody and haunting lyrics trigger my looking at the flower-covered hillside, the loveliness of which merges with the poignancy of the song. So many yellow daisies and purple Echinacea flowers! A great sense of hope and compassion. The world can be a good place! (Wednesday, July 22, 2015).

45. There is much construction on campus this summer, and a detour requires that I leave my usual sidewalk, cut across a parking lot, and use a stretch of sidewalk I've never traversed before. I am immediately touched by the atmosphere and mood of this path, which is tree-lined and shaded. This stretch of sidewalk is no more than 200 ft., but it has a strong sense of place. It is the trees that make this place special. There are many and one moves through a tree-canopied enclosure. The sidewalk meanders through the trees. The overall feeling is pleasant, comfortable, reassuring – quite wonderful! (Monday, July 20, 2015).

46. I'm walking down the hill of Leavenworth and note a boy in an orange shirt resting against the anchoring post of the stonewall at the corner. A young woman with a tiny terrier walks up Denison on the other side of that street. A blue truck stops at the corner, and the boy runs across the street to the truck. He must have been waiting for a ride. A car coming down Denison slows for the stopped truck. The boy gets in and the vehicle speeds away. I see irritation on the face of the driver, who obviously is not pleased that she's had to wait for the truck. The girl with the dog crosses Leavenworth and heads north on Denison. Odd how all these folks converge at that one corner and then disperse.

References

Allen, C. 2004. Merleau-Ponty's phenomenology and the body-in-space: Encounters of visually impaired children. *Environment and Planning D: Society and Space* 22: 719–735.

Behnke, E. 1997. Body. In *Encyclopedia of Phenomenology*, ed. L. Embree, 66–71. Dordrecht: Kluwer.

Bermudez, J.L., A. Marcel, and N. Eilan, eds. 1999. *The Body and the Self*. Cambridge, MA: MIT Press.

Carman, T. 2008. *Merleau-Ponty*. London: Routledge.

Casey, E. 1997. *The Fate of Place: A Philosophical History*. Berkeley: University of California Press.

———. 2009. *Getting Back into Place*. 2nd ed. Bloomington: Indiana University Press.

Cataldi, S.L. 2008. Affect and sensibility. In *Merleau-Ponty: Key Concepts*, ed. R. Diprose and J. Reynolds, 163–173. Stockfield: Acumen.

Cataldi, S.L., and W.S. Hamrick, eds. 2007. *Merleau-Ponty and Environmental Philosophy*. Albany: State University of New York Press.

Cerbone, D.R. 2006. *Understanding Phenomenology*. Durham: Acumen.

———. 2008. Perception. In *Merleau-Ponty: Key Concepts*, ed. R. Diprose and J. Reynolds, 121–131. Stockfield: Acumen.

Creswell, J.W. 2007. *Qualitative Inquiry and Research Design*. Thousand Oaks: Sage.

Diprose, R., and J. Reynolds, eds. 2008. *Merleau-Ponty: Key Concepts*. Stockfield: Acumen Publishing.

Donohoe, J. 2014. *Remembering Places*. New York: Lexington Books.

Evans, F. 2008. Chiasm and flesh. In *Merleau-Ponty: Key Concepts*, ed. R. Diprose and J. Reynolds, 184–193. Stockfield: Acumen.

Finlay, L. 2006. The body's disclosure in phenomenological research. *Qualitative Research in Psychology* 3: 19–30.

———. 2011. *Phenomenology for Therapists*. Oxford: Wiley-Blackwell.

Fullilove, M.T. 2004. *Root Shock*. New York: Ballantine Books.

Gallagher, S. 1986. Lived body and environment. *Research in Phenomenology* 16: 139–170.
García Márquez, G. 1967/1970. *One Hundred Years of Solitude*. New York: HarperCollins.
Heinämaa, S. 2012. The body. In *The Routledge Companion to Phenomenology*, ed. S. Luft and S. Overgaard, 222–232. London: Routledge.
Hill, M. 1985. Bound to the environment: Towards a phenomenology of sightlessness. In *Dwelling, Place and Environment: Toward a Phenomenology of Person and World*, ed. D. Seamon and R. Mugerauer, 99–111. Dordrecht: Martinus-Nijhoff.
Jacobs, J. 1961. *The Death and Life of Great American Cities*. New York: Vintage.
Jacobson, K. 2010. The experience of home and the space of citizenship. *The Southern Journal of Philosophy* 48 (3): 219–245.
Leder, D. 1990. *The Absent Body*. Chicago: University of Chicago Press.
Locke, P.M., and R. McCann. 2015. *Merleau-Ponty: Space, Place, Architecture*. Athens: Ohio State University Press.
Malpas, J. 1999. *Place and Experience*. Cambridge: Cambridge University Press.
———. 2006. *Heidegger's Topology*. Cambridge, MA: MIT Press.
———. 2009. Place and human being. *Environmental and Architectural Phenomenology* 20 (3): 19–23.
Merleau-Ponty, M. 1962. *Phenomenology of Perception*. New York: Humanities Press.
Moran, D. 2000. *Introduction to Phenomenology*. London: Routledge.
Morris, D. 2004. *The Sense of Space*. Albany: State University of New York Press.
———. 2008. Body. In *Merleau-Ponty: Key Concepts*, ed. R. Diprose and J. Reynolds, 111–120. Stockfield: Acumen.
Mugerauer, R. 1994. *Interpretations on Behalf of Place*. Albany: State University of New York Press.
———. 2008. *Heidegger and Homecoming*. Toronto: University of Toronto Press.
Oldenburg, R. 2001. *Celebrating the Third Place*. New York: Marlow & Co.
Pallasmaa, J. 2005. *The Eyes of the Skin: Architecture and the Senses*. London: Wiley.
———. 2009. *The Thinking Hand*. London: Wiley.
Pietersma, H. 1997. Maurice Merleau-Ponty. In *Encyclopedia of Phenomenology*, ed. L. Embree, 457–461. Dordrecht: Kluwer.
Relph, E. 1976. *Place and Placelessness*. London: Pion.
———. 2009. A pragmatic sense of place. *Environmental and Architectural Phenomenology* 20 (3): 24–31.
Romdenh-Romluc, K. 2012. Maurice Merleau-Ponty. In *The Routledge Companion to Phenomenology*, ed. S. Luft and S. Overgaard, 103–112. New York: Routledge.
Seamon, D. 1979. *A Geography of the Lifeworld*. New York: St. Martin's.
———. 2013. Lived bodies, place, and phenomenology. *Journal of Human Rights and the Environment* 4: 143–166.
———. 2014. Place attachment and phenomenology: The synergistic dynamism of place. In *Place Attachment: Advances in Theory, Methods and Research*, ed. L. Manzo and P. Devine-Wright, 11–22. New York: Routledge.
———. 2015. Situated cognition and the phenomenology of place: Lifeworld, environmental embodiment, and immersion-in-world. *Cognitive Processing* 16: S389–S392.
———. 2017. Hermeneutics and architecture: Buildings-in-themselves and interpretive trustworthiness. In *Hermeneutics, Space, and Place*, ed. B. Janz, 347–360. New York: Springer.
———. 2018. *Life Takes Place*. London: Routledge.
Seamon, D., and N. Gill. 2016. Qualitative approaches to environment-behavior research: Understanding environmental and place experiences, meanings, and actions. In *Research Methods for Environmental Psychology*, ed. R. Gifford, 115–135. New York: Wiley-Blackwell.
Seamon, D., and C. Nordin. 1980. Market place as place ballet: A Swedish example. *Landscape* 24 (October): 35–41.
Stefanovic, I.L. 2000. *Safeguarding Our Common Future*. Albany: State University of New York Press.

Toombs, S.K. 1995. The lived experience of disability. *Human Studies* 18: 9–23.
———. 2000. *Handbook of Phenomenology and Medicine*. Dordrecht: Kluwer Academic Publishers.
van Manen, M. 2014. *Phenomenology of Practice*. Walnut Creek: Left Coast Press.
Wachterhauser, B.R. 1996. Must we be what we say? Gadamer on truth in the human sciences. In *Hermeneutics and Modern Philosophy*, ed. B.R. Wachterhauser, 219–240. Albany: State University of New York Press.
Weiss, G. 2008. *Intertwinings: Interdisciplinary encounters with Merleau-Ponty*. Albany: State University of New York Press.
Weiss, G., and H.F. Haber, eds. 1999. *Perspectives on Embodiment*. New York: Routledge.
Willig, C. 2001. *Introducing Qualitative Research in Psychology*. Philadelphia: Open University Press.
Yardley, L. 2008. Demonstrating validity in qualitative psychology. In *Qualitative Psychology*, ed. J.A. Smith, 2nd ed., 235–251. Thousand Oaks: Sage.

Situating Interaction in Peripersonal and Extrapersonal Space: Empirical and Theoretical Perspectives

Shaun Gallagher

Abstract In this chapter I focus on the relationship between embodied intersubjective interactions and the kind of spaces that shape and are shaped by such interactions. After clarifying some of the theoretical background involved in questions about social cognition, I review several empirical studies that suggest that social interactions and social relations can change our perceptions of the reachable (peripersonal) space around us, as well as the more distant (extrapersonal) space beyond our immediate reach. These perceptions operate within the framework of material culture and impact our experience of space as it is organized by cultural artifacts and practices. In this respect, the analysis provided by Material Engagement Theory (Malafouris L, How things shape the mind. MIT Press, Cambridge, MA, 2013) helps us understand the role of material arrangements as they define affordances for action and interaction, correlated to transformations from individual body-schematic processes to intercorporeal processes in joint action. These same processes can be carried over into discussions of space and place as experienced on the larger stages of social-cultural activities.

Keywords Peripersonal space · Extrapersonal space · Body schema · Social cognition · Material engagement theory

1 Intersubjective Interaction and Social Cognition

Standard approaches to social cognition have little to say about the physical bodies of agents moving and interacting in space. The usual 'theory of mind' accounts, such as 'theory theory' or simulation theory, regard social cognition as an exercise in mindreading, understood as making inferences from folk-psychological

S. Gallagher (✉)
Philosophy, University of Memphis, Memphis, TN, USA

Faculty of Law, Humanities and the Arts, University of Wollongong, Wollongong, NSW, Australia
e-mail: s.gallagher@memphis.edu

T. Hünefeldt, A. Schlitte (eds.), *Situatedness and Place*, Contributions To Phenomenology 95, https://doi.org/10.1007/978-3-319-92937-8_5

knowledge, or running simulation routines where one imaginatively puts oneself in the other's situation in order to determine their mental states. These approaches tend to over-intellectualize processes involved in social interaction, and if they mention embodiment it tends to be a notion of body (or body-formatted) representations in the brain, e.g., the mirror neuron system (e.g., Goldman 2014; Goldman and de Vignemont 2009); the 'body in the brain' (Berlucchi and Aglioti 1997, 2010), rather than the body in the world.

An alternative view of social cognition, *interaction theory* (IT), draws on embodied, enactive, and extended conceptions of cognition. It appeals to the concepts of primary and secondary intersubjectivity found in developmental studies (e.g., Reddy 2008; Trevarthen 1979; Trevarthen and Hubley 1978). Primary intersubjectivity involves dynamical (sensory-motor) processes generated in face-to-face interaction – including bodily posture, movement, gesture, facial expression, vocal intonation, and so forth. Secondary intersubjectivity involves joint attention and joint action in pragmatic contexts. As we engage with others, we see, in their bodily comportment, including the kinematic details of their movement (Becchio et al. 2012), and in the particulars of the pragmatic situation, what they intend and what they feel. Meaning and emotional significance is co-constituted in the interaction, out in the world – not in the private confines of one or the other's head (De Jaegher et al. 2010).

From birth onward the infant is pulled into interactions with others. Although there are continuing debates about neonate imitation (Meltzoff and Moore 1977; see Coulon et al. 2013; Keven and Akins 2016; Oostenbroek et al. 2016; Vincini et al. 2017), whether such behavior is imitation, perceptual priming, contagion, or just arousal, it is relevant to early social development since the caregiver responds to the infants' spontaneous facial gesture with a further facial expression, which in turn provokes more behavior in the infant. In this way, the infant is drawn into intersubjective interactions. Indeed, a caregiver may over-interpret the infant's response as a social response, a phenomenon that simply promotes further interaction (Vincini et al. 2017). In primary intersubjectivity, then, there is continuous interaction with caregivers from birth. At 6 months infants start to perceive grasping as goal directed; at 10–11 months infants are able to parse some kinds of continuous action according to intentional boundaries (Baldwin and Baird 2001; Baird and Baldwin 2001; Woodward and Sommerville 2000). Infants start to perceive various movements of the head, the mouth, the hands, and more general body movements as meaningful, goal-directed movements (Senju et al. 2006). This is not mindreading, but rather interaction combined with perception of emotional, embodied meaning in the postures, movements, gestures, facial expressions, and actions of others. Accordingly, the mind is not hidden or private; rather, it is given in the other person's embodied comportment so that appeal to unperceived hidden mental states is not required for most of our everyday interactions.

In this respect, if mirror neurons (MNs) are involved in such processes, it's not clear that they are simulating or matching the perceived action of the other person. Empirical studies suggest that there is no precise matching in MN activation as one would require in a simulationist interpretation (Catmur et al. 2007; Csibra 2005;

Dinstein et al. 2008). In contrast, IT defends an enactivist interpretation of MN activation. Specifically, MN activation is regarded as part of a response to the other's action and may involve motor preparation for one's continued response to the other.

2 Action and Interaction Spaces

When one looks for empirical support for this enactivist interpretation, one is led to studies that involve the distinction between peri- and extra-personal space. Caggiano et al. (2009), for example, showed differential activation of MNs in premotor cortex of rhesus monkeys for peripersonal space *versus* extrapersonal space.

> A portion of these spatially selective mirror neurons... encode space in operational terms, changing their properties according to the possibility that the monkey [may be able to] interact with the object [and agent]. These results suggest that a set of mirror neurons encodes the observed motor acts not only for action understanding, but also to analyze such acts in terms of features that are relevant to generating appropriate... behavioral responses to those actions.... mirror neurons not only may represent a neuronal substrate for understanding 'what others are doing', but also may contribute toward selecting 'how I might interact with them'. (Caggiano et al. 2009, p. 403)

In other words, if the other person or the object they have an interest in is within peripersonal space so that I can directly interact with the person and/or object, MN activation is different than if the person and/or object is out of reach in extrapersonal space. Bonini et al. (2014) extended this finding to include canonical neurons (i.e., neurons that activate in response to either using or simply seeing a manipulable object) and canonical-mirror neurons (which combine the functions of MNs and canonical neurons) (see Fig. 1).

Further clarification about intersubjective interaction and our perception of social space is provided by Heed et al. (2010), who demonstrated that multisensory integration is modulated 'only if the partner performs a task in the participant's peripersonal space', and not if the task is performed in extrapersonal space. Likewise, Teneggi et al. (2013) have shown that the presence of others in extrapersonal space shapes the experience of peripersonal space. Peripersonal space is constricted when faced with another individual standing in extrapersonal space, as compared to when we are facing a mannequin placed at the same location. The peripersonal boundary also changes as a function of the social experience we are having with the individual facing us (see Fig. 2).

The same study showed that following an economic game, peripersonal space boundaries between self and other tend to merge, but only if the other person behaved cooperatively during the game. Peripersonal boundaries are sensitive not only to the presence of others but also to interactions with others – and by valuation of other people's behavior during the interaction.

> [C]onsistent with approaches to cognition suggesting that mental processes are situated and embodied in our physical experiences.... high-level social and cognitive representations

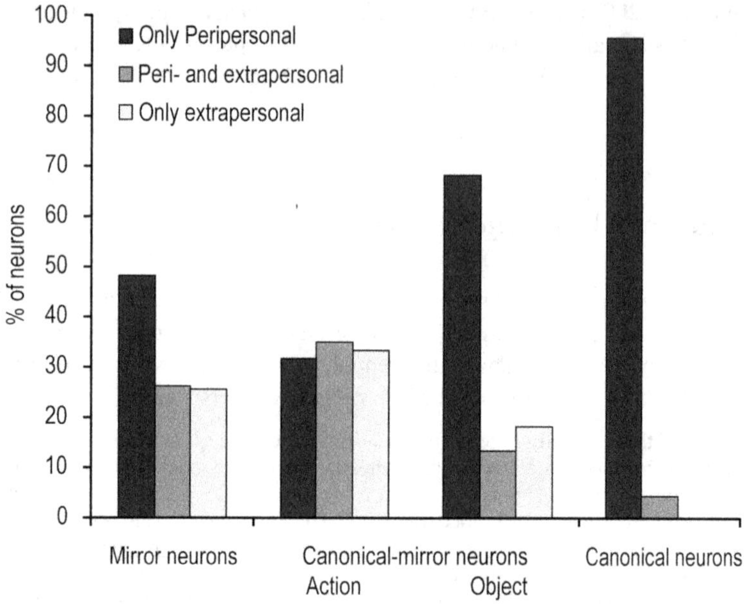

Fig. 1 Histograms showing the relative proportion of mirror, canonical-mirror, and canonical neurons selectively responding to visual stimuli presented either in the peripersonal (black) or extrapersonal (white) space, or activated for stimuli presented in both space sectors (gray). (From Bonini et al. 2014)

> (e.g., cooperation [in economic games]) are immersed or recoded into the physical and perceptual experiences of the body, thereby providing concrete and rich feelings that facilitate prediction, evaluation, and social behavior. (Teneggi et al. 2013, p. 409).

3 Space Perception and the Socially Extended Mind

Building on the concepts of distributed cognition and the extended mind (e.g., Menary 2010), Malafouris (2013) has developed Material Engagement Theory (MET), which proposes that *things* in the surrounding environment – artifacts, tools, instruments, etc. – and the very materiality of such stuff – shape our experiences and our instrumental and social practices (also see Gosden 2008). For example, Malafouris shows how artifacts can change our cognitive practices and ecologies, including our lived space, "reshaping a hybrid space that intersects bodily, peripersonal, and extrapersonal areas" (Malafouris 2013, p. 245).

Consistent with MET, there is good empirical evidence for changes in the experience of peripersonal space (and its underlying neural basis) during tool use (Fig. 3). This extension of peripersonal space is sometimes described as tool incorporation into the body schema (Berti and Frassinetti 2000; Farnè et al. 2005; Maravita and Iriki 2004). Maravita and Iriki (2004) show bimodal neurons in the parietal cortex

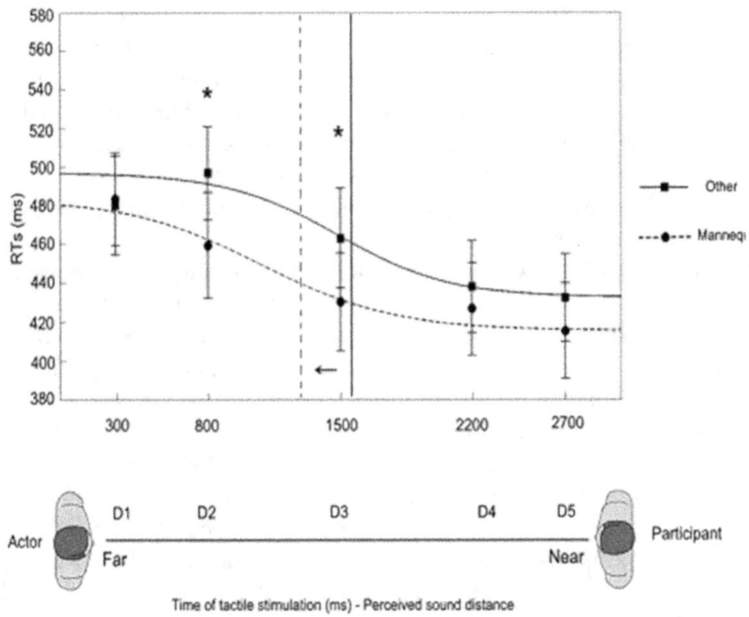

Fig. 2 Peripersonal space measured indirectly via reaction times to perceived sounds: higher reaction times for other *versus* mannequin. (From Heed et al. 2010)

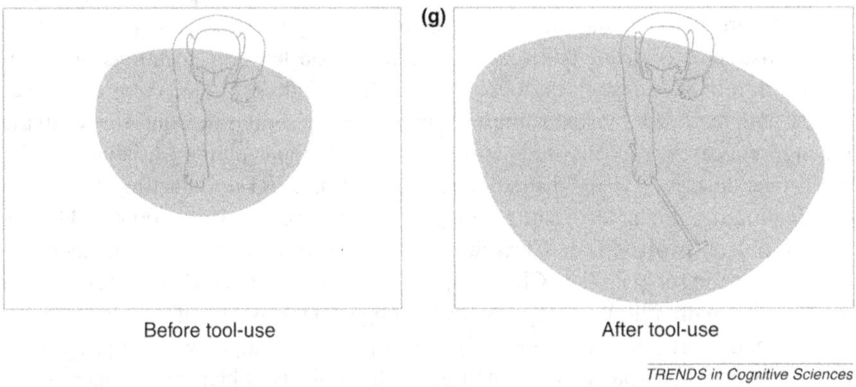

Fig. 3 Expanded peripersonal space during tool use. (From Maravita and Iriki 2004)

have both somatosensory and visual receptive fields focused on the hand. Neuronal activation increases for the visual receptive field as one looks near one's hand. After tool use, the visual receptive field expands to include the effector end of the tool even if that end is outside of peripersonal space.

Beyond tool use, we can think of larger scale phenomena concerning social ecologies and cultural niches. Consider the notion of the *socially* extended mind (Gallagher 2013) – the idea that larger structures, both physical structures (built environments, architectures), and social institutional structures (and practices) shape experience, cognition, and intersubjective relations. No doubt there is significant research on the effect of architectural structure and design on our experience of space (e.g., Pasqualini et al. 2013). But it's not just architectural structure and instrumental manipulation (as in tool use) that modulate how we experience things in near or far space. Cultural institutions and practices do something similar. Consider two brief examples that help to show the effect of religious practices on the experience of space.

Gallagher et al. (2015) were able to replicate in simulated virtual and mixed reality environments feelings of awe and wonder experienced by astronauts during space travel. In experiencing earth from the perspective of outer space, there was not only a shift in spatial perspective, as might be expected, but also significant *scale effects,* e.g., the feeling of being very small in contrast to the vastness of space, that also significantly involved affective experience. This study found significant differences in the EEG alpha range correlated to phenomenological reports of specific categories of awe and wonder experiences and expressed shifts in spatial perspective. Follow-up questionnaires showed that awe experiences were less frequent among those subjects who engaged in more religious practices. Those who engaged in less self-reported religious practices had more first-order experiences of awe involving changes in spatial perspective as well as scale effects connected with experiences of the vastness of outer space, than those who reported more religious practice.

In a second example, Michiel van Elk (2014) used the crossmodal congruency effect (CCE), a direct measure of multisensory integration in peripersonal space. The measure involves introducing an incongruency in spatial perception using a visual distractor on trials that require discrimination between locations of vibrotactile stimuli. Response rates (to tactile stimuli) are quicker in the congruent cases (when both tactile and visual stimuli are presented on same side), and slower in the incongruent cases. Van Elk used pictures of artifacts representing Christian, Hindu, or profane objects as visual distractors. Reaction times (RTs), reflecting spatial discrimination (up-down, left-right within peripersonal space), were modulated by the type of object involved (i.e., Christian, Hindu, or profane) and the religious background of the participant (i.e., Christian, Hindu, or non-religious). The study showed longer RTs with religious objects; even longer ones with unfamiliar religious objects, and shorter RTs with non-religious objects. This study was complicated by the inclusion of graspable *versus* non-graspable objects, which may constrain the results. van Elk's interpretation acknowledged the role of physical affordances but also suggested there were clear cultural influences on experiences of peripersonal *versus* extrapersonal space.

Related to this conclusion, given differences in conceptions about self as more independent *versus* interdependent in North Americans *versus* East Asians, respectively (Kitayama and Park 2010), and differences in perception itself between more foreground-focused Europeans and North Americans *versus* more background aware East Asians (Nisbett and Miyamoto 2005), one could think that there may be differences in the experience of peri- and extrapersonal space as well. Soliman and

Glenberg (Soliman and Glenberg 2014; Soliman et al. 2013) show a difference in spatial judgment related to just such cultural differences. In estimating distance to a member of one's in-group (in extrapersonal space) the more *interdependent* the participant, the shorter the estimated distance to the in-group confederate, and particularly for shorter estimates over longer distances (e.g., 20 m *versus* 7 m). For example, for in-group and out-group comparisons, Arabs perceiving Arabs replicated the original in-group result; Non-Arabs perceiving Arabs (1) estimated they were closer; and (2) reversed the (self-identified) independent vs interdependent measures. Soliman and Glenberg (2014) argue that culture is embodied and has an effect on our perception of distance. They suggest that embodiment allows for "the possibility of redefining interdependence and independence by recourse, not to explicit values and beliefs, but to sensorimotor interpersonal repertoires that are either narrowly tuned to in-group interactional practices, or more broadly tuned, and appearing to be more culturally diverse" (p. 214).

This research suggests a broader conception of affordances (Gibson 1977) – not just physical and pragmatic action possibilities, but social and cultural interaction possibilities. This is consistent with a point made by Adams et al. (2004, p. 343):

> The independent constructions of self and relationship that are prominent in mainstream American settings are not merely beliefs about separation. Instead, they are linked to a reality of separation that is built into structures of everyday life like dating practices, residence in apartment units, and individual ownership. Similarly, the more relational or interdependent constructions of self and relationship that are prominent cultural patterns in many West African settings are not merely beliefs about connection. Instead, they are linked to a reality of connection that is built into structures of everyday life like arranged marriage, residence in lineage compounds and the practice of eating meals from a communal bowl.

As Soliman and Glenberg (2014) suggest, social practices that involve differences between sleeping patterns for infants, the ways that care-givers carry around their infants, or whether and to what extent they allow the child to crawl, may affect their sense of peri- *versus* extrapersonal and spatial perception. The point here is that our sense of peripersonal *versus* extrapersonal space, and variations in spatial perception, may reflect not only physical and pragmatic affordances (e.g., graspability), but also social and cultural affordances. Ramstead et al. (2016, p. 3) define two kinds of cultural affordances. The first type, which they call "natural" cultural affordances, consist of possibilities for action that depend on the agent exploiting correlations in the environment *via* abilities that are culturally derived. Such affordances are closely tied to material artifacts and the built environment. The second type, "conventional affordances" are "possibilities for action, the engagement with which depends on agents' skillfully leveraging explicit or implicit expectations, norms, conventions, and cooperative social practices".

In this regard the notion of affordance space or affordance landscape is a useful conceptual tool. We can define changes in affordance space (Brincker 2014; Gallagher 2015) or the landscape of affordances (Rietveld and Kiverstein 2014) as changes in the range of possibilities provided by any active movement in body or change in (physical, social, cultural) environment. An individual's occurrent affordance space is defined by *evolution/embodiment* (the fact that she has hands, for instance), *development* (her life-stage – infant, adult, aged), and by *social and cultural practices* (normative constraints) – all of which enable and constrain the individual's action possibilities.

4 Extension of Peripersonal Space During Joint Action

Malafouris (2013) applies Material Engagement Theory (MET) to individual action. MET can also be extended to include various forms of joint action – where two or more people interact in order to accomplish a particular goal (Gallagher and Ransom 2016). Joint actions may expand or shrink the affordance landscape relative to what individuals may accomplish on their own. This likewise can affect space perception and the boundaries of peri- and extrapersonal space.

Soliman and Glenberg (2014) studied changes in the perception of peripersonal space during joint action. Two participants coordinated for 5 min moving a flexible wire back and forth to cut through a candle, requiring close coordination to keep the wire tight. Using complex visual *versus* tactile interference measures that demonstrate incorporation of tools and extension of peripersonal space during tool use,[1] they showed that peripersonal space expands to include the space around the other participant. The expansion of peripersonal space during tool use is often interpreted to mean the extension of the body schema to incorporate the tool. Accordingly, Soliman and Glenberg refer to the case of a social dynamics that expands peripersonal space to include the space around the other participant as the establishment of a joint body schema (also see Soliman et al. 2015).

The notion of a joint body schema and the extension of peripersonal space during joint action helps cash out Merleau-Ponty's notion of intercorporeity. Merleau-Ponty considers intercorporeity to be a pre-reflective, relational phenomenon that may be even more basic than intersubjectivity. He sketches this notion in his early work, *The Phenomenology of Perception.* "There is, between … [my] phenomenal body, and the other person's phenomenal body such as I see it from the outside, an internal relation that makes the other person appear as the completion of the system"

[1] Soliman and Glenberg (2014) used a complicated technique called the 'flash/buzz' paradigm (Maravita et al. 2002) to indirectly measure to what degree peripersonal space is expanded. A buzzer is attached to the index finger and another attached to the thumb. One of the buzzers is activated, and the task is to indicate (using a foot pedal) whether the thumb or index finger had been stimulated. Two LED lights are located next to the 2 fingers, respectively. LEDs flash in temporal synchrony with the buzzer but spatially incongruously (e.g. the thumb is stimulated by the buzzer but the LED next to the index finger is flashed). In these incongruous trials, reaction times are slowed in reporting which finger is buzzed, and participants make errors. Maravita et al. (2002) show that tool use can modify the interference on the incongruous trials. Before using a tool, the LEDs interfere with localization of the vibration only when they are near the fingers, that is, in peripersonal space. After using a tool (e.g., a rake), the LEDs can be located at the end of the tool, outside of the usual peripersonal space, and still interfere with localization of the active buzzer. This is taken as a measure of how much the body schema has incorporated the tool (or how much peripersonal space has expanded).

(Merleau-Ponty 2012, p. 368). This concept is directly related to the spatiality of the body and the notion of reversibility. Merleau-Ponty's favorite example of reversiblility, taken from Husserl (1989), is that of one person's two hands touching each other. If I use my right hand to touch my left hand, there is the immediate possibility of a reversibility – that my right hand touching can immediately become the touched; and my left hand touched can immediately become the touching. If the touching-touched is in some sense simultaneous, in terms of our single-minded attention it is not, but involves a dynamic sequential reversibility, not unlike the reversing of the Necker cube in vision, but one that can be done at will (Merleau-Ponty 1968, p. 141). Merleau-Ponty's enigmatic example, "Reversibility: the finger of the glove that is turned inside out" (Merleau-Ponty 1968, p. 317), links reversibility to the fact that hands are incongruent counterparts or what geometers call enantiomorphs (Kant 1992; see Morris 2010; Gallagher 2006), which is easily demonstrated with gloves. One cannot put a left-hand glove on one's right hand, for example. But if one turns the left-hand glove inside out, it takes the shape of a right-hand glove. The reversibility, in this case with respect to shape, is clearly sequential; although the glove simultaneously has the incongruent shapes in its structure, they can only manifest themselves (as the touching-touched phenomenon manifests itself in experience) in sequence.

Merleau-Ponty extends the concept of reversibility to the case of one person touching another person's hand: "when touching the hand of another, would I not touch in it the same power to espouse the things that I have touched in my own?" (Merleau-Ponty 1968, p. 141). Merleau-Ponty suggests that in the case of my own two hands touching "there exists a very peculiar relation from one to the other, across the corporeal space – like that holding between my two eyes – making of my hands one sole organ of experience" (Merleau-Ponty 1968, p. 141). Yet he wants to maintain that something like this can exist between two bodies. "Why would not the synergy exist among different organisms, if it is possible within each? Their landscapes interweave, their actions and their passions fit together exactly" (Merleau-Ponty 1968, p. 142). The empirical studies of the joint body schema lend support to Merleau-Ponty's phenomenological analysis.

The affordance landscape defined by joint-action possibilities is not only dependent on the individuals involved (the spatiality of their bodies, their skill levels, strengths, weaknesses, etc.), but on the material aspects of things, their spatial arrangements, changes in the perception of peri- and extrapersonal spaces involved in actions and joint actions, affected by the specific motoric aspects of such activities, but also by the normative aspects of social and cultural practices. Joint action may expand the affordance landscape or it may limit my individual affordances – but this can depend on all of the variables just mentioned – something we can also think about on larger scales of institutional arrangements.

5 Instituted Spaces

The affordance landscape for joint action may be externally constrained (intentionally or unintentionally) by institutional arrangements and practices – and this includes the arrangement of space and our built environments. One example is the use of architectural 'brutalism' on university campuses in the 1960–1970s. Whether the use of this architecture was intended to quell student protests or not, that may have been one of many effects (Jerry 2004; Temple 2014). We design places to facilitate, to inhibit or to shape joint actions, rituals, social practices, institutional procedures – all of which can change our affordance landscapes, expanding or limiting our action and interaction possibilities. As the architect Louis H. Sullivan (2013, p. 282) pictures it, architecture is meant to "vitalize building materials, to animate them with a thought, a state of feeling, to charge them with a social significance and value, to make them a visible part of the social fabric, to infuse into them the true life of the people". We can think of various formal institutions that design practices and correlated places that can, like material buildings, expand or contract affordance landscapes, animate or de-animate them.

The practices of science, religions, legal institutions, universities, and so forth, require, or at least motivate the design of appropriate spaces. Mark Johnson suggests that, in regard to architecture and embodied action, "we live in and through our ongoing interactions with environments that are both physical and cultural. The structures we make are loosely adapted to the functions we perform […] [We] order our environments to enhance meaning in our lives and to open up possibilities for deepened and enriched experience" (Johnson 2015, p. 33). Architecture, understood as a kind of knowledge or hermeneutics, is the study of an order that we bring into existence, "not only as a building, but as a human situation, in the space-time of experience" (Pérez-Gómez 1998, p. 24; see Gallagher et al. 2017).

Consider more ad hoc arrangements, as found in circumstances that recently defined the Arab Spring or the 'Occupy' movements, where governments were discouraging people from gathering together in the same space to organize and resist (Gallagher and Ransom 2016). Despite this prohibition, coordinated collective activity was made possible, not only through the use of social media, but in the medium of public space. The form that such activity could take was in large part enabled and constrained by the detailed affordances of material culture – the technology, the buildings, the places directly tied to cultural practices – allowing for particular kinds of social-political action that would have been impossible in the absence of such materially constituted affordances. The same kinds of material and social affordances define the types of actions and sanctions that are utilized by government agencies in response to such protests. Not only cultural history, but these new actions and counter actions made certain places more important.

6 Conclusion

I've focused on the relationship between embodied intersubjective interactions and the way that such interactions shape and are shaped by the way we experience space. Empirical studies support the idea that social interactions and social relations can change our perceptions of peripersonal and extrapersonal space. Such perceptions operate within the framework of material culture, which impacts our experience of space as it is organized by cultural artifacts and practices. Material spatial arrangements help to define affordances for action and interaction, correlated to transformations of individual body-schematic processes and intercorporeal processes in joint action. I've suggested that these same processes can be carried over into discussions of space and place as experienced on the larger stages of social-cultural activities.

Acknowledgements I presented an earlier version of this paper as a keynote lecture at the ICSC 2015: *6th International Conference on Spatial Cognition*. Rome 7–11 September 2015. My research for this project has been supported by the Humboldt Foundation's Anneliese Maier Research Award, and as a Senior Visiting Researcher at Keble College, Oxford in 2016.

References

Adams, G., S.L. Anderson, and J.K. Adonu. 2004. The cultural grounding of closeness and intimacy. In *Handbook of Closeness and Intimacy*, ed. D.J. Mashek and A. Aron, 321–339. New York: Psychology Press.

Baird, J.A., and D.A. Baldwin. 2001. Making sense of human behavior: Action parsing and intentional inference. In *Intentions and Intentionality: Foundations of Social Cognition*, ed. B.F. Malle, L.J. Moses, and D.A. Baldwin, 193–206. Cambridge, MA: MIT Press.

Baldwin, D.A., and J.A. Baird. 2001. Discerning intentions in dynamic human action. *Trends in Cognitive Science* 5 (4): 171–178.

Becchio, C., V. Manera, L. Sartori, A. Cavallo, and U. Castiello. 2012. Grasping intentions: From thought experiments to empirical evidence. *Frontiers in Human Neuroscience* 6: 117.

Berlucchi, G., and S. Aglioti. 1997. The body in the brain: Neural bases of corporeal awareness. *Trends in Neuroscience* 20: 560–564.

———. 2010. The body in the brain revisited. *Experimental Brain Research* 200: 25–35.

Berti, A., and F. Frassinetti. 2000. When far becomes near: Remapping of space by tool use. *Journal of Cognitive Neuroscience* 12 (3): 415–420.

Bonini, L., M. Maranesi, A. Livi, L. Fogassi, and G. Rizzolatti. 2014. Space-dependent representation of objects and other's action in monkey ventral premotor grasping neurons. *The Journal of Neuroscience* 34 (11): 4108–4119.

Brincker, M. 2014. Navigating beyond "here & now" affordances – On sensorimotor maturation and "false belief" performance. *Frontiers in Psychology* 5: 1433. https://doi.org/10.3389/fpsyg.2014.01433.

Caggiano, V., L. Fogassi, G. Rizzolatti, P. Thier, and A. Casile. 2009. Mirror neurons differentially encode the peripersonal and extrapersonal space of monkeys. *Science* 324: 403–406.

Catmur, C., V. Walsh, and C. Heyes. 2007. Sensorimotor learning configures the human mirror system. *Current Biology* 17: 1527–1531.

Coulon, M., C. Hemimou, and A. Streri. 2013. Effects of seeing and hearing vowels on neonatal facial imitation. *Infancy* 18 (5): 782–796.
Csibra, G. 2005. Mirror neurons and action observation. Is simulation involved? *ESF Interdisciplines*. http://www.interdisciplines.org/mirror/papers/.
De Jaegher, H., E.A. Di Paolo, and S. Gallagher. 2010. Can social interaction constitute social cognition? *Trends in Cognitive Sciences* 14 (10): 441–447.
Dinstein, I., C. Thomas, M. Behrmann, and D.J. Heeger. 2008. A mirror up to nature. *Current Biology* 18 (1): R13–R18.
Farnè, A., A. Iriki, and E. Làdavas. 2005. Shaping multisensory action-space with tools: Evidence from patients with cross-modal extinction. *Neuropsychologia* 43: 238–248.
Gallagher, S. 2006. The intrinsic spatial frame of reference. In *The Blackwell Companion to Phenomenology and Existentialism*, ed. H. Dreyfus and M. Wrathall, 346–355. Oxford: Blackwell.
———. 2013. The socially extended mind. *Cognitive Systems Research* 25–26: 4–12.
———. 2015. Doing the math: Calculating the role of evolution and enculturation in the origins of mathematical reasoning. *Progress in Biophysics and Molecular Biology* 119: 341–346.
Gallagher, S., and T. Ransom. 2016. Artifacting minds: Material engagement theory and joint action. In *Embodiment in Evolution and Culture*, ed. C. Tewes, 337–351. Berlin: de Gruyter.
Gallagher, S., L. Reinerman, B. Janz, P. Bockelman, and J. Trempler. 2015. *A Neurophenomenology of Awe and Wonder: Towards a Non-reductionist Cognitive Science*. London: Palgrave Macmillan.
Gallagher, S., S. Martínez Muñoz, and M. Gastelum. 2017. Action-space and time: Towards an enactive hermeneutics. In *Hermeneutics: place and space*, ed. B. Janz, 83–96. Berlin: Springer.
Gibson, J.J. 1977. The theory of affordances. In *Perceiving, Acting, and Knowing: Toward an Ecological Psychology*, ed. R. Shaw and J. Bransford, 67–82. Hillsdale: Lawrence Erlbaum.
Goldman, A.I. 2014. The bodily formats approach to embodied cognition. In *Current Controversies in Philosophy of Mind*, ed. U. Kriegel, 91–108. New York/London: Routledge.
Goldman, A.I., and F. de Vignemont. 2009. Is social cognition embodied? *Trends in Cognitive Sciences* 13 (4): 154–159.
Gosden, C. 2008. Social ontologies. *Philosophical Transactions of the Royal Society of London. Series B* 363: 2003–2010.
Heed, T., B. Habets, N. Sebanz, and G. Knoblich. 2010. Others' actions reduce crossmodal integration in peripersonal space. *Current Biology* 20 (15): 1345–1349.
Husserl, E. 1989. *Ideas Pertaining to a Pure Phenomenology and to a Phenomenological Philosophy: Second Book, Studies in the Phenomenology of Constitution*. Trans. R. Rojcewicz and A. Schuwer. Boston: Kluwer Academic Publishers.
Jerry, R.H. 2004. A brief exploration of space: Some observations on law school architecture. *University of Toledo Law Review* 36: 85–93.
Johnson, M. 2015. The embodied meaning of architecture. In *Mind in Architecture: Neuroscience, Embodiment, and the Future of Design*, ed. S. Robinson, J. Pallasmaa, and J., 33–50. Cambridge, MA: MIT Press.
Kant, I. 1992. Concerning the ultimate ground of the differentiation of directions in space. In *The Cambridge Edition of the Works of Immanuel Kant. Theoretical Philosophy, 1755–1770*, ed. D. Walford and R. Meerbote, 365–372. Cambridge: Cambridge University Press.
Keven, N., and K.A. Akins. 2016, July. Neonatal imitation in context: Sensory-motor development in the perinatal period. *Behavioral and Brain Sciences* 14: 1–107. https://doi.org/10.1017/S0140525X16000911.
Kitayama, S., and J. Park. 2010. Cultural neuroscience of the self: Understanding the social grounding of the brain. *Social Cognitive Affective Neuroscience* 5 (2–3): 111–129.
Malafouris, L. 2013. *How Things Shape the Mind*. Cambridge, MA: MIT Press.
Maravita, A., and A. Iriki. 2004. Tools for the body (schema). *Trends in Cognitive Sciences* 8 (2): 79–86.

Maravita, A., C. Spence, S. Kennett, and J. Driver. 2002. Tool-use changes multimodal spatial interactions between vision and touch in normal humans. *Cognition* 83 (2): B25–B34.
Meltzoff, A., and M.K. Moore. 1977. Imitation of facial and manual gestures by human neonates. *Science, New Series* 198 (4312): 75–78.
Menary, R., ed. 2010. *The Extended Mind*. Cambridge, MA: MIT Press.
Merleau-Ponty, M. 1968. *The Visible and the Invisible: Followed by Working Notes*. Trans. A. Lingis. Evanston: Northwestern University Press.
———. 2012. *Phenomenology of Perception*. Trans: D.A. Landes. New York: Routledge.
Morris, D. 2010. The enigma of reversibility and the genesis of sense in Merleau-Ponty. *Continental Philosophy Review* 43 (2): 141–165.
Nisbett, R.E., and Y. Miyamoto. 2005. The influence of culture: Holistic versus analytic perception. *Trends in Cognitive Sciences* 9 (10): 467–473.
Oostenbroek, J., T. Suddendorf, M. Nielsen, J. Redshaw, S. Kennedy-Costantini, J. Davis, and V. Slaughter. 2016. Comprehensive longitudinal study challenges the existence of neonatal imitation in humans. *Current Biology* 26 (10): 1334–1338.
Pasqualini, I., J. Llobera, and O. Blanke. 2013. "Seeing" and "feeling" architecture: How bodily self-consciousness alters architectonic experience and affects the perception of interiors. *Frontiers in Psychology* 4: 354.
Pérez-Gómez, A. 1998. The case for hermeneutics as architectural discourse. In *Architecture and Teaching-Epistemological Foundations. 31st EAAE Workshop (European Association for Architectural Education)*, ed. H. Dunin-Woyseth and H., 21–29. Paris: Comportements.
Ramstead, M.J.D., S.P. Veissière, and L.J. Kirmayer. 2016. Cultural affordances: Scaffolding local worlds through shared intentionality and regimes of attention. *Frontiers in Psychology* 7: 1090.
Reddy, V. 2008. *How Infants Know Minds*. Cambridge, MA: Harvard University Press.
Rietveld, E., and J. Kiverstein. 2014. A rich landscape of affordances. *Ecological Psychology* 26 (4): 325–352.
Senju, A., M.H. Johnson, and G. Csibra. 2006. The development and neural basis of referential gaze perception. *Social Neuroscience* 1 (3–4): 220–234.
Soliman, T., and A.M. Glenberg. 2014. The embodiment of culture. In *The Routledge Handbook of Embodied Cognition*, ed. L. Shapiro, 207–219. New York: Routledge.
Soliman, T., A. Gibson, and A.M. Glenberg. 2013. Sensory motor mechanisms unify psychology: The embodiment of culture. *Frontiers in Psychology* 4: 885.
Soliman, T.M., R. Ferguson, M.S. Dexheimer, and A.M. Glenberg. 2015. Consequences of joint action: Entanglement with your partner. *Journal of Experimental Psychology: General* 144 (4): 873–888.
Sullivan, L.H. 2013. *Kindergarten chats and other writings*. Toronto: Reed Books, Inc.
Temple, P. 2014. *The Physical University: Contours of Space and Place in Higher Education*. London: Routledge.
Teneggi, C., E. Canzoneri, G. di Pellegrino, and A. Serino. 2013. Social modulation of peripersonal space boundaries. *Current Biology* 23: 406–411. https://doi.org/10.1016/j.cub.2013.01.043.
Trevarthen, C. 1979. Communication and cooperation in early infancy: A description of primary intersubjectivity. In *Before Speech*, ed. M. Bullowa, 321–347. Cambridge: Cambridge University Press.
Trevarthen, C., and P. Hubley. 1978. Secondary intersubjectivity: Confidence, confiding and acts of meaning in the first year. In *Action, Gesture and Symbol: The Emergence of Language*, ed. A. Lock, 183–229. London: Academic.
van Elk, M. 2014. The effect of manipulability and religion on the multisensory integration of objects in peripersonal space. *Cognitive Neuroscience* 5 (1): 36–44.
Vincini, S., J. Yuna, E.H. Buder, and S. Gallagher. 2017. Neonatal imitation: Theory, experimental design and significance for the field of social cognition. *Frontiers in Psychology – Cognitive Science* 8: 1323. https://doi.org/10.3389/fpsyg.2017.01323.
Woodward, A.L., and J.A. Sommerville. 2000. Twelve-month-old infants interpret action in context. *Psychological Science* 11 (1): 73–77.

Spatial Conception of Activities: Settings, Identity, and Felt Experience

William J. Clancey

Abstract The "situated" perspective in the analysis and design of socio-technical systems reveals how people conceive of activities as social-interactive *settings*, thus relating situated cognition to themes in the philosophy of place. In the socio-technical approach to developing technology, social scientists and computer scientists ground designs of automation (e.g., software, devices, vehicles) in ethnographic studies of how people interact with each other, tools and representations (e.g., computer displays), and their environment (e.g., facilities). These studies characterize people's *activities* – how they conceive of what they are doing in particular settings. A person's understanding of activities that a physical location affords for certain socio-technical purposes – the spatial conception of activities – makes a location a meaningful place. Technical knowledge and methods ("what I am doing now"), personal identity ("who I am being now"), and behavior norms (e.g., "how well I am doing now") co-develop in the activity setting. This paper elucidates the multi-dimensional physical, conceptual, and interactive nature of settings with examples from ethnographic studies of robotically mediated field science on Mars and analog expeditions on Earth. The felt and aesthetic experience of being in places that are Mars-like and working on Mars itself further reveals the emotional aspect of cognition that motivates and orients scientific work, exploration, and associated artistic expressions.

Keywords Activity · Spatial conception · Setting · Situated cognition · Identity · Ethnography · Work practice · Planetary science · Analog environment · Socio-technical design · Aesthetic experience

W. J. Clancey (✉)
Florida Institute for Human & Machine Cognition, Pensacola, FL, USA
e-mail: wclancey@ihmc.us

T. Hünefeldt, A. Schlitte (eds.), *Situatedness and Place*, Contributions To Phenomenology 95, https://doi.org/10.1007/978-3-319-92937-8_6

1 Introduction

What is situated cognition and how does it relate to "place"? In this paper, I discuss the notions of "situatedness" and "place" in socio-technical studies intended to inform the design of artificial intelligence technology in software, devices, vehicles, and robots that help people do their work. These analyses, often called "workplace studies," (Anderson 1994, 1997; Harper 2000; Luff et al. 2000) focus on the logistics and circumstances of work – what people do, where, when, with whom, using what tools, etc. – that enables the routine workflow of tasks and information, called *work practice*. In contrast, other methods for designing work systems (e.g., "business process modeling" and "task analysis") more abstractly describe the *functions* people accomplish, often called *tasks*, modeling work as transformations of data and materials in the manner of a manufacturing assembly-line process. Studies of practice are about the *behaviors and beliefs of people* who at times are "at work." Workplace studies focus on how people interact with each other and their environment, including communicating, manipulating materials, using devices and vehicles, gesturing, and moving.

Work practice can be modeled as patterned behaviors called *activities*, which occur over time in some *setting* (Clancey 2002). The activity unit of analysis relates psychological and social-interactive analyses of people, settings, and activities, and thus reveals *how reasoning is "situated"* – that is, it occurs as experiences and interactive behaviors, rather than only being a mental process in the brain. Activity models capture circumstantial, logistic details that cognitive models ignore: where reasoning (inferring, calculating, imagining) occurs, where data and information are located (e.g., paper documents, graphics, displays), and under what circumstances reading and writing information occurs. This cognitive science research topic is broadly called *situated cognition* (Lave 1988; Clancey 1997, 2008). Models of activities holistically relate roles, procedures, records, automated tools, facilities, schedules, etc. In contextualizing work functions in this manner, work practice studies reveal temporal and spatial constraints on perception, comprehension, and decision-making, in addition to how impasses are resolved and incidental communications and assistance, all of which is highly relevant to the design of work systems (e.g., Clancey et al. 2013). Or turned the other way, applying the "situated" perspective to analysis and design of socio-technical systems reveals how people conceive of activities spatially, relating situated cognition to themes in the philosophy of place.

Over the past 30 years the socio-technical design approach to developing technology has matured as a partnership among social scientists and computer scientists (e.g., Greenbaum and Kyng 1991; Nardi 1996; Salvador et al. 1999; Rönkkö 2010). Consequently, the research methodology combines ethnographic observation (Spradley 1980; Clancey 2006a), social theories of practice and cognition (Leont'ev 1979; Wertsch 1979), and often computer models and simulations, such as simulating work practice in Brahms (Clancey et al. 1998). Socio-technical design projects typically seek to automate work that is tedious, error prone, and costly, and may involve developing tools such as the mobile robotic laboratories on Mars that enable work that otherwise would be too dangerous or impossible to otherwise perform (Clancey 2012).

This paper considers developing tools for working on Mars either for future astronauts or for planetary scientists operating robotic physical surrogates remotely from Earth or Mars orbit. In these design projects, field scientists use the term "place" routinely in their work to refer to named geographic regions on a map (e.g., Meridiani Planum on Mars, Devon Island in the Canadian Arctic). In an activity-based analysis, the term "setting" refers to a *conceptualization*, how people in a given culture think about a place, including norms for who can be there, how we might dress and behave, and the suitability and adaptation of a place for certain activities (Barker 1968). In general, the notion of "setting" plays a central role in ethnographic studies for practical design projects.[1] Viewed through the metaphor of a theatrical play, the place is the stage where we experience a constructed world of rooms, decor, and props, constituting the setting, in which a story unfolds revealing the relationships among people, technology, and their environment.

In effect, a key purpose of this paper is to elucidate the physical, mental/emotional, interactive, and social nature of settings, based on ethnographic studies of field science on Mars and expeditions on Earth in places that physically resemble Mars (analog environments). In particular, I illustrate how people engaged in activities conceive of settings in terms of *spatial relations of physical geography* (e.g., a hill), *activities*, *social relationships*, and *things* (e.g., tents). These examples reveal that people's conception of activity and setting blends with their identity, and the mutual relations of action and meaning emerge in practice in an actor's conceptualization of "what I am doing (here) now" (Clancey 1997; Wenger 1998).

How we conceive of settings can reveal both the nature of conceptualization as well as how social, bodily, and activity notions pervade our concept of places. A fundamental notion that I wish to make clear throughout is that "situations" are *conceptual,* that is mental constructs, and our conception of activities relates many categories and feelings – social, logical, sensory, kinesthetic, and aesthetic. A "situation" for an actor is not only a physical place and what is perceived. Rather "the situation" is the actor's encompassing understanding at a given time of roles, responsibilities, setting, methods of interacting with the environment and other people (norms), tools, etc. that are guiding and giving meaning to his/her ongoing activities. Similarly, when we say an activity is "social" we do not mean that the actor is necessarily physically present with other people, but rather the activity is *conceptually social* – the manner in which it is done, what needs to be done, how it will be

[1] We can view "setting" in workplace studies as an *etic* term (a structural, conceptual distinction of the people in the culture being studied, which may or may not be in their lexicon), and "place" as an *emic* term (an analytic, culturally general distinction, named and often graphically related to other places as in a map). In human culture the place–setting distinction blends; for example, constructed (e.g., buildings, plazas) and natural places (e.g., lakes, mountains) alike are known as both physical places and cultural settings for activities. Of particular interest are settings in which only certain people are allowed to participate and activities are circumscribed, such as the settings related to Mars field science discussed in this chapter.

evaluated, etc. are affected by the actor's understanding of the norms, methods, knowledge, expectations, etc. of the social group in which the actor is participating, whether it be an organization, group of friends, or family members. Such conceptions are often tacit, implicit in learned relations that are not necessarily ever reflected upon and articulated in descriptions about the self, world, and others. They constitute the "know how" of practice.

As I will show by several examples at the end, the situated cognition perspective also provides a way to characterize our experience of the "aesthetics of a place," namely as a spatial conception that is both an action-orientation (physical) and a feeling (non-verbal experience). In my analysis, *conceiving* is inherently emotional, though this aspect has often been omitted entirely from cognitive models and artificial intelligence (Damasio 1994). One of my objectives in this paper is to provide an empirical, practice-based perspective on settings that others may perhaps analyze further to provide philosophical distinctions useful for design projects.

This chapter begins with examples of settings and practices during Haughton-Mars analog expeditions in the Canadian High Arctic, illustrating the methods and analytic eyeglasses of socio-technical design. I then elaborate the story of these expeditions to include the notion of identity and explain the notion of activity with examples from a simulated mission in the Flashline Mars Arctic Research Station. This analysis is complemented by discussing how field science expeditions are possible on Mars today without being physically present: the Mars Exploration Rovers are not only physical surrogates but collaborative tools, designed to fit the social, embodied nature of human cognition. The final analytic section, on the felt and aesthetic experience of being in places that are Mars-like and working on Mars itself, emphasizes the emotional aspect of cognition that motivates and orients scientific work, exploration, and associated artistic expressions. This chapter concludes with a brief review of perspectives on urban and architectural design in the 1950s to 1970s that has strongly influenced socio-technical studies of workplaces by relating the physical, pragmatic-interactive, social, and felt dimensions of activity settings. Finally, the themes that organized the lectures and conference workshop in which the ideas in this chapter were first presented are revisited from the perspective of situated cognition, specifically in the context of workplace studies.

2 Settings and Practices in Haughton-Mars Expeditions

The Haughton-Mars Project (HMP, Lee and Osinski, 2005) is a multidisciplinary group that carries out scientific and engineering expeditions near Haughton Crater on Devon Island in the Canadian Arctic (Fig. 1). The crater and its environs constitute a "Mars analog" because the island is undeveloped, almost entirely barren and without animals. The landscape is mostly breccia, a kind of rubble stone caused by the meteor impact, and mud, with ravines and hills shaped by a glacier that once covered it. Field scientists, chiefly geologists and biologists applying the method of comparative planetology, study this place to understand the geology and possibility

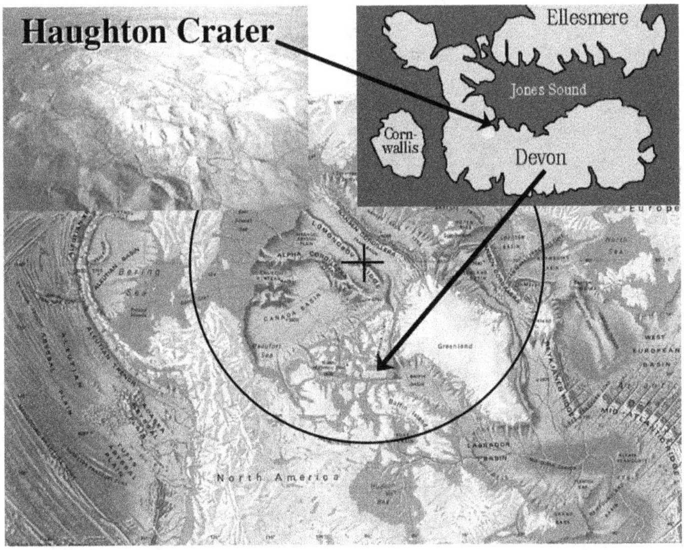

Fig. 1 Geographic location of Haughton-Mars Expedition

of life on Mars, past, present, or future. For example, the "valley networks" seen on Devon Island are known to have formed under a glacier, and thus it is hypothesized that similar morphologies on Mars were formed in a similar way (e.g., thus explaining how water erosion could be possible despite Mars' frigid temperature and thin atmosphere).

But Haughton Crater also constitutes a Mars analog setting of another sort. Insofar as it resembles Mars in the imagination and is devoid of every distraction of civilization, the place can be used for an activity called a "Mars analog mission." In particular, the participants of the expedition living in base camp or members of the Mars Society living in the nearby Flashline Mars Arctic Research Station (Zubrin 2003) can pretend that they actually are on Mars and constrain their activities accordingly, such as always wearing a simulated spacesuit outdoors. Thus, the environs of the Haughton Crater become a kind of stage setting for role playing, in which they come to understand what a Mars mission requires, developing roles within the crew, experimenting with tools such as drills, exploring and documenting the region, communicating with a remote "mission support" team, and so on. Thus, it should be apparent that in this activity of a "Mars mission simulation," the behaviors of the members are constrained by their *conception of the place as being Mars* – in their imagination, in the layout of the camp and their habitat; and through their actions they create and reinforce their understanding that "this is what being on Mars could be like."

My ethnographic research in the HMP began in 1998 when we were camped along the Haughton River (Clancey 2001). My original objective was to describe the people and their roles, the setting, and activities of the expedition, particularly to elucidate the nature of exploration, a central topic for NASA. At that time, both

Fig. 2 ATV parking area near work tents (Credit: NASA Haughton–Mars Project, William J. Clancey)

cognitive and social studies of science focused on laboratory settings; indeed, information-processing psychological models reduced "discovery" to mental problem-solving processes, such as inferring equations to relate numeric data (Langley et al. 1987; Downes 1990). I wanted to show how people explored a place and how this activity related to scientific discovery.

One of my own first discoveries was how people parked the all-terrain vehicles (ATVs), orienting the vehicles' front to the power cord that ran from a generator by the river to the work and mess tents (Fig. 2). In an ethnographic study it is useful to remind ourselves that whatever we observe could have been different. Nobody directed the group to park in this way. I never heard anyone correct someone for parking elsewhere. Yet there was plenty of open space and no important reason for lining the ATVs in this way. In seemed to me that the cord and how it was used resembled a curb in a street back on Earth. We say that the cord "provided a resource" for ordering people's parking behavior.

Parking ATVs in this manner emerged entirely at this place and time during a few weeks in this particular 1998 expedition; it became the practice of the group. This practice can be interpreted as having both pragmatic and socially important aspects. The ATVs are shared and so parking them by personal tents would be inconvenient to others and appear to claim ownership. Parking the ATVs near the shared tents makes them ready-at-hand for their use, which was almost always organized and planned with the group leader. We can also see in the photo that this emergent organization is exploited in providing a convenient place to sit outside while talking to someone; this is a good place to meet and interact, just as other members of the expedition pause to talk in the space created between the work and mess tents.

In summary, the organization of the shared tents, power cord, and ATVs on this river plateau emerged in practice as a setting for the expedition; it is a place socially ordered by activities with different levels of abstraction and purposes, which we might name, "carrying out an expedition," "laying out base camp," "parking ATVs," "sitting/standing and talking." We can observe directly in this setting that practices develop from mimicked, recurrent actions – individuals independently create patterns (the line of ATVs) that others perceive, prompting their own actions that confirm and strengthen the pattern (e.g., driving up and parking alongside an ATV). Thus, practice and setting emerge through precedent – every individual action provides a context for future action and reinforces the group's patterning of "what you can do here," "how we use this place." Such norms develop and are reinforced in activity.

Tying this back to the socio-technical design perspective: Just as the members of the expedition were not instructed about how to park ATVs and do not even necessarily reflect on what they are doing, a robotic ATV, to fit the expeditions practices, would need to discover and follow the patterns in the behavior of people and other robotic ATVs. The robotic system must regulate its behaviors according to precedent, to follow what other actors are doing; in particular, the conception of functional use of space ("what you can do here") is common to the members of the group and robotic actors must be capable of detecting and adapting to this conception as it develops in practice. This is not to preclude explicit instruction or discussion, but simply to note that not all behavior needs to be planned and verbally expressed in procedures; it can be learned through activity. Ryle's (1949) distinction of knowing how and knowing that is fundamental to making robots or software programs ("agents") into social actors.

Figure 3 provides a broader perspective revealing other social relations that constitute the HMP-99 expedition setting. Here we see that personal tents, where individuals sleep and store their belongings, are aligned along the edge of the river

Fig. 3 Layout of the tents at the Haughton-Mars Expedition 1998 (Credit: NASA Haughton–Mars Project, William J. Clancey)

plateau. I recall asking the group leader where I could place my tent (the blue tent on the right); he suggested to put it near his (on the far right). The organization of personal tents is not haphazard. Again it illustrates that members of the group seek, respect, and create patterns, and indeed that visible order is important in the HMP-99 community of practice. The use of the place, this river plateau, also respects visible physical boundaries; the shared white tents and ATVs are more or less in the middle, the personal tents as far away as practical, but separated for some privacy. The yellow-orange tent in the foreground is the latrine – far enough away to respect its special purpose, but still conveniently close. Another white wall tent on the left is a kind of office for the expedition commander; its separation shows that it is not shared and open for improvised use like the work and mess tents. In short, common ideas of private and public spaces and different functions in designing homes and offices are visible in the layout of the HMP base camp. But the expedition, by being remote from existing facilities, roads, etc., makes salient how shared experience and recurrent behaviors develop to exploit, interpret, and adapt a physical place to create settings for different personal, joint, and improvised activities.

Looking inside the work tent (Fig. 4) we can discern more patterns that a robotic system would need to know. To begin, the term "work tent" doesn't completely characterize how this place is used. Again, functions emerged in practice, which is to say that the shared understanding of this setting was constructed through individual actions, with precedents establishing a conceptual (and visible) norm, that are reinforced or adapted by subsequent actions. People have dedicated work areas on tables whose boundaries are loosely marked (e.g., the upright brief case in the upper-left photo) and are not always negotiable. Other aspects of the social order include not sitting in or moving someone else's chair; mostly working alone and quietly; storing large personal items under the table; not removing someone else's charging device; and dedicating one table for storing exploration tools, etc. As needs develop, certain adaptations might require mentioning what you are doing, such as borrowing an extension cord; the request makes note that you are aware of the norm and provides an opportunity for objection or a suggested alternative. A bipedal robotic participant would need to know all this.

As is well known, design of workspaces like this tent benefits from being flexible so the organization and purposes of the place can develop in practice. Here the expedition leader provided tables and power outlets; the rest was constructed by the group, with the layout and practices constraining the people who arrived in subsequent "rotations" of the expedition. Understanding emergent order has been fundamental to the notion of "human-centered design." Traditional work systems design (e.g., of facilities and software tools) typically began with "functions" (idealized roles for people), "requirements" (e.g., a priori description of procedures and protocols), a design (often involving automation), interfaces (i.e., means for people to interact with the facilities and tools), and finally training for the workers in using the provided workspace and tools. In contrast human-centered design involves the workers as co-designers, provides flexibility for adaptation (along with structure to prompt uses), observes and learns from practice, and then iterates again with the workers to improve the work system design (roles, procedures, tools, furniture, etc.).

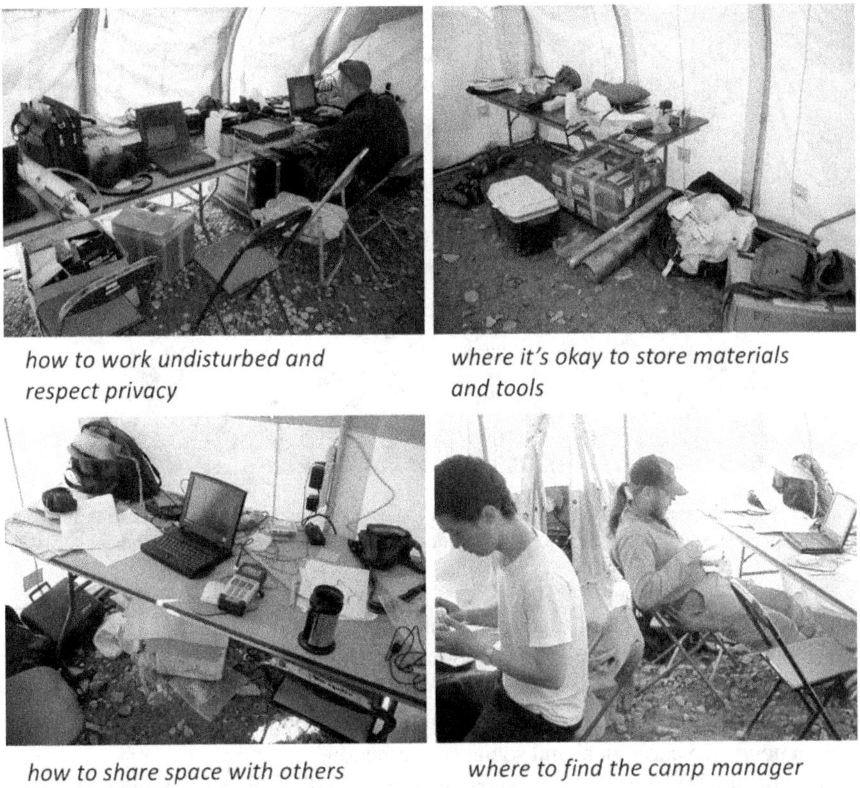

Fig. 4 Ethnographic observations of a "Work Tent" – what a robotic assistant would need to know (Credit: NASA Haughton–Mars Project, William J. Clancey)

These principles are applicable whether designing a robotic laboratory on Mars, a workplace office building (or a Mars habitat), a self-driving car, or a voice-commanded "agent" on a mobile device. In each case, the people, activities, and settings will develop in practice. Or to pick up this dynamic by one thread pertaining to this book's theme – how a place becomes a setting for people engaged in activities is itself a situated activity, involving different levels of pragmatic interactions, reflection, and often deliberate redesign experiments.

3 The Relation of Identity to Activities and Settings in Mars Expeditions

The notion of identity provides an essential link between psychological and social analyses of people (Clancey 1999, Chapter 1): on the one hand, the neuropsychological nature of consciousness in coordinating perception, conceptual

Fig. 5 People express their identity as members of expeditions and space exploration missions through patches and clothing. Explorers stand on hilltops to mark their accomplishment

understanding, and action enables people to be social actors (i.e., the physical "mechanism" that enables learning and intelligent behavior); on the other, a person conceives "what I am doing now" in terms of his/her roles and activities in a group (i.e., the conceptual "content" that constitutes practical knowledge). Identity is thus both a neuro-psychological and social phenomenon.

Identity at HMP and in the Mars Society is reified by patches and blazoned across our chests (Fig. 5), a distinctly human way of expressing membership. This impromptu picture of the author pokes fun at the iconic photographs of mountain-climbing explorers, but it also signifies something real about the pride and joy of *being here, in this place, at this moment.* In a playful way, this little pile of rubble symbolizes a mountain towering above the crater, and standing tall on this high point claims a degree of personal accomplishment in the expedition. Put another way, recording participation in the expedition through this photograph expresses an identity, which conceptually blends the social fact of membership with an emotion and aesthetic perception of the place.

During Mars mission simulations in an analog environment, we are designing and experimenting with habitats, tools, and practices for surviving and exploring in a harsh landscape (Fig. 6). The future-oriented interest of computer scientists in developing technology for Mars is enhanced and supports the interest of planetary scientists in investigating and understanding Mars. We are learning how we might live and work on Mars by co-constructing activities and settings, as well as our identities as space-faring explorers. By role playing Mars missions in a simulated habitat in an extreme environment we can use our imaginations to design and experiment with technology, operational procedures, roles, schedules, space suits, vehicles, and facilities we might use on Mars (Clancey 2006b). We include remote support teams, exploration traverses of the landscape, logs and records, and so on.

Fig. 6 Flashline Mars Arctic Research Station (FMARS), a setting for Mars mission simulations constructed by the Mars Society on the rim of Haughton Crater (Credit: William J. Clancey)

We create a simulated Mars setting, such as placing the habitat on the edge looking out over the crater, to prime our creativity, and through the physical and social interactions of activities inside and outside the habitat, we learn what works and discover new problems. It is a place for creative play, in which we pretend we really are crew members on Mars. The ethnographic "participant observation" method of observing and recording thus naturally involves contributing to the expedition's scientific and technical activities, adopting the identity of an expedition member. This identity is in large part earned by being an analyst and designer who helps realize the vision of exploring Mars.

The relation of activities to settings is fundamental and merits elaboration. To begin, an "activity," like a "task," is a unit of analysis, a way of interpreting, "chunking," and labeling human behavior. A key part of an ethnographic study is categorizing what people are doing, where, when, with whom, how, etc. as activities. One of the common activities in FMARS is working alone together (Fig. 7). The individuals are quietly working on their own projects. They are not collaborating, but cooperating, sharing a common resource (the hab space) and respectful of each other's need for silence to concentrate.

Activities are consciously choreographed behaviors in the world (Clancey 2002). Participating in this group requires regulating one's behavior by an understanding "What am I doing now?" that blends both "Who am I being now?" and "How am I behaving now?" That is, besides attending to the substance of one's personal activity (e.g., the Discovery Channel photographer is reading a novel; the biologist to his left is reading a scientific article), one must attend to the implicit norms of this group, such as not making many outloud remarks to share interesting ideas. To be a social actor, the personal and social aspects must be both immediate and always present in each person's understanding of the activity. That is, their conceptualiza-

Fig. 7 Working alone together. A group activity that is not collaborative – realizes common activity. FMARS, July 2000 (Credit: William J. Clancey)

tion is coordinating multiple levels or dimensions of their behavior simultaneously, involving their choice of what to do at this time, their manner of doing their work, and even how they are sitting. Notably, the photographer and biologist could have worked in their staterooms just behind their chairs, with the doors closed. Instead, they have chosen to be near the two people working at the table, angled as if to show their conceptual orientation of being part of the group, forming a unit.

More generally, an actor's ongoing conceptualization of an activity determines roles, communications, attention, schedule, problem-solving methods, tools, etc., that are appropriate for the given setting. For the analyst, understanding how the actors conceive "the situation" enables detecting and appreciating how they are interacting with each other and their environment. During joint activities such a meeting or preparing meals, the group's interactions involve different *modes of interaction* such as monitoring, cooperating, waiting, not interfering, and face-saving. Even sitting quietly is a form or mode of interaction. In short, a situated cognition perspective enables analyst–designers to view this floor of the FMARS habitat as a social-interactive place, a setting that prompts, reinforces, and is constructed by individual actions. The norms are visible in how the place is used (i.e., behaviors), how it becomes a stage for socially meaningful activities (Clancey et al. 2005).

As we saw in the camp and work tent layouts, the patterns of behavior in the FMARS habitat are emergent, tacit, and orderly. The crew members express their identity by respecting and reconstructing these norms moment by moment. Each person knows which chairs can be used (those at the table are shared) and where they are expected to be placed; but these patterns don't need to be discussed or regulated by rules unless a breakdown occurs ("Do you know where my chair is?").

Fig. 8 Two activities: exploring and working (or resting) at your workstation. FMARS, July 2001 (Credit: William J. Clancey)

Figure 8 shows two examples that vividly illustrate how what we do in different places is conceptually regulated. To be sure, riding around on ATVs in the Arctic wearing white canvas suits with garbage can lids on your head is quite a feat of the imagination! Here imagination and creativity have produced an activity that is very far from civilized life, yet seems perfectly natural when role playing "being a member of a crew on Mars." The landscape, weather, vehicles, and clothing mutually constitute the setting, afforded by this unmarked, unbounded place for driving and exploration. Our behavior is mediated by how we are thinking about our activity, so the context is conceptual, coordinating our perception and actions – "What I am doing now" is imagined and enacted, meanings and possibilities for social interaction in this environment feeding each other.

Simulated missions provide an ideal opportunity to study the scientific practice of exploration in gathering data about, categorizing, and theorizing about a place; understanding this activity is fundamental to the design of mobile robotic laboratories for field work on other planets. Notice also that "exploring" has an open quality that a task analysis would need to gloss, as the opportunistic processes of poking around, probing with instruments, making notes, etc. is not well expressed by a fixed, step-wise procedure. Rather, the activity of exploring is interactive and circumstantial, an emergent product of the physical geography, perception, tools, interests/expertise, and time available. Explorers are always re-evaluating "What is the quality of the work I am doing now?" "What are the opportunities (or challenges) posed by this place for my exploration interests?" And socially orienting to the scientific community: "Am I gathering sufficient and appropriate data for presentations and publications?" When the various factors harmonize, interest increases, and the activity focuses, such that the place becomes a setting for perhaps prolonged work of applying instruments, sampling, data recording, and systematic survey that a task analysis can describe well. In paleontology and archaeology, which share characteristics of planetary field science, the setting will be formalized in a visible shared coordinate system and every found item (bone or shard) labeled and recorded in a database. People are assigned places to excavate and slowly the place becomes an ancient historical setting for reconstructed events (animals collected in the bend of a muddy river; rooms of homes in a village). Understanding the morphologies of Mars' landscape in terms of historical climatology is similar.

To elaborate the psychological and social aspect of activities and identities, consider two crew members working on their computers in the shared FMARS workstation area (Fig. 8, right). While Vladimir is awake and working, Katy has taken a moment to rest. If we were to rouse her and ask, "What are you doing now?" She might say, "I'm tired, I'm contemplating." "Why don't you rest in your stateroom?" "I'm not done yet; I'm still working on my email." "What do you need to do?" "I'm a geophysics graduate student at MIT, I have ongoing projects back in Cambridge where I live." Pressed further by a journalist perhaps, she might add, "I'm an American just visiting here in Canada as part of the FMARS crew." Crucially, each of these descriptions is simultaneously true and each characterizes what she is doing now and who she is being now. Her identity as a crew member, a student, and a citizen are blended and may affect how she carries out FMARS activities. For example, a journalist may seek to highlight her story because she is a woman–scientist, or she might volunteer to be responsible for certain exploration activities in a role that complements her Ph.D. research.

Consider how robotic systems and agent assistants as they are currently constructed lack such histories of nested activities and identities. We might model activities and give roles and responsibilities to computer systems, but still the robot/agent must observe, learn, and adapt because practices are prone to change. Furthermore, the social license for adaptation is challenging to formalize: Why can't a worker at a telephone call center rest with her head against the desk, although it is perfectly allowable in FMARS? What social understandings do we tacitly share that makes obvious that norms are radically different in "being at work in an office" and "being a crew member in a Mars habitat"? How are common self-regulatory needs such as resting accomplished in different settings?

Bringing this back to the theme of the philosophy of place, a setting for a person is always both personal and social, both physical and conceptual, and often both visually perceived and verbally named. Settings are bound to our identity, who we are, the activities in which we participate. An end-of-mission portrait of the FMARS 2001 Rotation 2 crew (Fig. 9) shows how each person has a "sense of place," what is theirs, where they belong, and each know (and can see) that they share this understanding and the effect is cooperative (I'll respect your privacy knowing you will respect mine). The photograph also shows a more basic level of coordinated action – they have chosen to stand by their rooms and are mimicking each other in holding onto a door or the wall, signifying that they know and possess their place. This plays out in much more sophisticated ways in carrying out the simulated mission, planning and participating in exploration and reporting, sharing in chores, and contributing to hab life by cooking for and entertaining each other.

To recap, the Mars habitat examples illustrate how people make places into settings for carrying out activities; spatial layout of things and how they sit, stand, relax, etc. makes visible how they are conceiving social relations. Thus their *motives*, which are expressed in the activity (e.g., expressing "pride of place," napping,

Fig. 9 Group portrait of FMARS 2000 rotation 2 crew (Credit: William J. Clancey)

concentrated reading), are in a triadic relation with *norms* and the *affordances of things and places* (e.g., the edge of the stateroom doors affords gripping with the hand).[2] This analytic, situated cognition perspective – relating conceptually coordinated, embodied behavior; social identity and practices; and the physical environment – is useful for design of buildings, furniture, work processes, and technology.

To round out the FMARS habitat example, particularly to illustrate technology design, we need to consider collaborative work. An important example is the activity of exploring a place remotely using a mobile robotic laboratory, such as the Mars Exploration Rovers (MERs). On Mars MER-like physical surrogates might be operated from habitats, orbiting space stations, or from Earth as they are today in NASA's missions. Socio-technical studies of MER missions reveal an important aspect of spatial conception, namely people's ability to imaginatively project themselves into the body of the rover, visualizing and simulating what they would do if they were standing there on Mars (Clancey 2012; Fig. 10).

David Des Marais, an astrobiologist, described the MER scientists' experience this way:

> The first few months of the mission, they had these huge charts on the wall, engineering drawings of the rover, with all of the dimensions. We'd have some geometric question, "Well can we see this; can we reach this? Is this rock going to be in shade or is it in the sun?" We'd go stand and stare at those charts, and over time we stopped doing it so much because we began to gain a sense of the body. That's definitely projecting yourself into the rover. It's just an amazing capability of the human mind – that you can sort of retool yourself. (Clancey 2012, p. 108)

[2] Affordances are dynamic relations among a physical place/thing and the intentions and behavioral capabilities of actors; they are not properties that reside in either the place/thing or the actor. Other examples: the upper bunk of the stateroom affords sleeping and being used as a standing desk; the table affords seating for six people at meals or four people using computers.

Fig. 10 Mars Exploration Rover (MER) missions named Spirit and Opportunity, operated from Jet Propulsion Lab, Pasadena, California. Clockwise from upper-left: Schematic of MER instruments; MER science team informal discussion; pointing to a feature in the life-size panorama on the table; photograph taken on Mars of MER instrument deployment arm with sensors, microimager, and rock abrasion tool (RAT) (Credits: NASA/JPL; Chin Seah; NASA/JPL; William J. Clancey)

Thus, the operation of the MER rover illustrates one of the central tenets of situated cognition, *embodiment* – in a nutshell, we are capable of tacitly (non-verbally) conceiving spatial relations that relate to how we have interacted and moved within our environment. Learned gestures and coordinated actions are coupled to concepts, ranging from simple actions like throwing something to hit a moving object to named motions, such as a "do–si–do" in a square dance. Thus, conceptualization appears to be a higher-order form of categorization (incorporating perception and motion patterns) for coordinating behavior (Clancey 1999).

Doing field science on Mars requires providing people with a means of *experiencing presence*, of being in that place. The technologies developed include MER's stereo panorama camera, use of "virtual reality" displays and eyeglasses at JPL, life-size shared photographs, and even the proportions of the MER (e.g., the panoramic camera is 1.5 m high) and configuration of the instruments (e.g., the RAT is on the end of the arm, and can be positioned on a rock like a hammer).

Crucially, the design of MER transforms the surface of Mars into a field science setting, a place where the scientists and engineers together can carry out an expedition, sensing, sampling, photographing, manipulating soil with the wheels, and in the case of the MER named "Opportunity," moving over dozens of kilometers over more than a decade. The scientists' spatial conception of their activity – the setting for their field science exploration – ranges from close-up photographs of grains of sand to wide vistas in craters and over the plains of Mars. But this conception includes as well nearby places they haven't seen and visual imagination prompted by photographs from orbit of routes they might yet or could have followed. By the very nature of exploration, a field science expedition necessarily includes seeking out and understanding "places we can go that we don't know much about." Indeed, for many years the Mars Science Laboratory team was learning and hypothesizing about possible interesting and tractable routes within Gale Crater up the central mountain, which rises 5.5 km above the valley floor. Thus the setting for their exploration activity was constructed from the physical affordances, their interests, and the capabilities of the rover and its instruments.

In short, technology of mobile robotic laboratories enables the activity of doing field science on Mars by facilitating, in the scientists' and engineers' imagination, the spatial conceptualization of being on the surface of Mars in the body of the rover – experiencing its capabilities and limitations for analyzing and traversing the unfolding terrain. Put another way, the process of exploring Mars is necessarily contextual and *the place becomes a setting, a stage for action,* by virtue of the conceptual embodiment that mental processes and the robotic laboratory's design enable. Further, the design of the rover, the software tools for planning and programming its actions, the layout of the rooms and even the floors of the JPL building itself (with ongoing missions in different locations on Mars sharing engineering facilities), and of course the roles, work practices, and schedules of the teams all fit together to constitute a *work system* that physically, intellectually, and socially organizes and makes this activity of exploring Mars possible.

4 The Felt and Aesthetic Experience of Places

We have considered how Mars expeditions occur in special places, both remote Mars analog locations on Earth and of course on Mars, referring particularly to the practice of scientific work. But we would be greatly remiss if we didn't mention the aesthetic, emotional motivations for exploration, which are arguably an essential personal motivation for being a scientist (Clancey 2012, chapter 9). Photographs from Devon Island and Mars reveal a great deal about our capability to project ourselves – conceiving of places spatially such that they are settings for activities – and the feelings that accompany this conception. Figure 5 illustrates feelings and emotion with respect to identity; here we consider photographs that are themselves considered to be aesthetically pleasing, that is, what you experience when looking over a landscape (or seeing it as if you are physically there).

During HMP expeditions the group sometimes traveled by ATV late in the summer evening, when the sun was low on the horizon, to a hilltop overlooking Haughton Crater (Fig. 11, left). Some people called this "a spirit place," as it evoked feelings of something sacred, transcending everyday settings, distant and foreign,

Fig. 11 The Hills, "A Spirit Place" (Credit: NASA Haughton–Mars Project, William J. Clancey) and Meridiani Planum viewed from Opportunity Mars Exploration Rover (Credit: NASA/JPL)

yet visible and apparently reachable. This is an unlived-in world, a place revered by the Inuit, not partitioned, named, protected, or trammeled. It is a place quite unlike where we live and work; it opens for us like the curtains revealing an anticipated play, a yet unknown stage with not yet imagined actions, unpeopled by actors and artifacts, pristine, like land created before man, and abiding today alone.

The emotional aspect of perception is a relatively new topic in cognitive science. For example, Keltner and Haidt (2003) conducted experiments demonstrating that "awe is a very powerful trigger of prosocial behavior…. [One is] "Less focused on the self and more in tune with the present moment" (quoted by Mikulak 2015, p. 18). The very choice of Haughton Crater and its environs as an analog site was prompted by its awe-inspiring landscapes, photographs of which motivated others to join the expedition; and this effectively promoted collaborative research that transformed the place further into a setting for Mars mission simulations (Clancey 2014). Its relevance to comparative planetology (an extrinsic, systemic value) and the emotionally evocative feeling (an intrinsic, personal value) stimulate our imagination of being on Mars and confer an identity that "I am realizing the dream." As De Botton (2002, p. 183) says: "And insofar as we travel in search of beauty, works of art may in small ways start to influence where we would like to travel *to*."

The view of Meridiani Planum (Fig. 11, right) reveals a different aspect of Mars, namely that it is not only an unvisited, unblemished world, but a place that nobody owns. Thus as a setting it has an idealized character; although a place we have traveled through and gotten to know, it still exists as if in a spatial ether, apart from the human body, indeed a place only for our spirit. Through the rovers we exist on Mars through a physical surrogate and conceptually – in our scientific understanding, as the story of Mars; and in future imagined activities, as the story of humanity. Thus Mars is a place conceived scientifically, aesthetically, and emotionally, a setting par excellence, visually experienced like a vivid dream, known as if it is directly real to us, as if we have actually traveled through Meridiani Planum, Gusev and Gale craters, buoyed in awe with opportunity, spirit, and curiosity.

Fig. 12 "The Ravine," Haughton Crater, Devon Island (Credit: NASA Haughton–Mars Project, William J. Clancey); Crater tracks of Curiosity, Mars Science Laboratory (Credit: NASA/JPL)

Two related photographs from the Arctic and Mars reveal how spatial conceptions are affordances for activities, that is, the perception of the place is coupled to a conception of and primes possible movements, transforming it into a setting, a place where an activity could be carried out. Exploring the crater in 1998 by ATV, we came upon a ravine (Fig. 12, left) and paused.

The ravine's perspective suggested turning off our planned route, to drive up the rubble, perhaps to gain a better view of the hills above and what lay beyond. Places are thus "seen through activities," conceived as possibilities for action that satisfies needs – for passage, shelter, reconnaissance, storage, cultivation, etc. As expressed by Fox (2000, p. 42), "When defining our relation to landscape spatially, the temporal side of the equation has to be taken into account…." The ravine beckoned for exploration, it physically afforded an ATV path, and its receding perspective urged me bodily forward. Already engaged in our simulated Mars "traversal" on an afternoon away from base camp, I felt a strong desire to go and be *up there*. But our small group turned away, continuing forward, so this place has always seemed like someone met in a crowd whom I would have liked to know. It was more mysterious knowing that almost certainly no human being had ever tread there – it was not even yet a "road less taken." Viewed askance, it looks and feels like Mars on Earth. It is one of uncountable places in the solar system that nobody knows, yet we know exist and might become a setting for our exploring, "a place for being," or in the case of Mars, a setting for a future civilization.

Another photograph, taken in Gale Crater on Mars (Fig. 12, right), looks back on the tracks of the Curiosity rover through some small hills through which it passed.[3] Examining this image, notice how we move into a place with our eyes, conceiving where we can go, what we might touch: "Our attention…engages in motion over the landscape" (Dewey 2014). This tacit conception lends aesthetic value, pleasing-ness to the view.

[3]The image has been white-balanced to show what the Martian surface materials would look like if under the sunlight of Earth's sky.

Conceiving is a reconstructive process of remembering (Bartlett 1932/1977; Clancey 1997). Thus, "what I am seeing here" is seen through past places, perhaps how we *felt*, how we were oriented in activities when in places physically like this, and what activity the place affords. Visual perception (including in the imagination) is not necessarily accompanied by speech, but a felt experience, an *apprehension*, constructed from how categories of multiple dimensions – visual, verbal, auditory, olfactory, kinesthetic – were related in prior constructions. Emotion organizes and reinforces conceptualization (and remembering) – thus conceptualizing is bound to *attitude*, a personal stance towards that aspect of the experienced world (Bartlett 1932/1977). Therefore, conceptualization of activity–settings is always accompanied by some feeling/attitude, such as a sense of security (protective or threatening), movement (past, present, or future), or awe (including wonder, mysterious feeling, and admiration of beauty).

By virtue of being reconstructive from embodied experience, visual perception and conceptualizing a place are also *self-referential* in the sense of relating to our bodies and our actions. Figure 13 provides a striking example. Some viewers see footprints here, quite reminiscent of the iconic boot prints of Apollo astronauts; in fact, they are MER's tire tread marks, created by deliberate attempts to "excavate" the soil by removing the top layer of sand and dust. As an apprehension, a non-discursive conception, the sight of the boot prints concomitantly projects ourselves into the image, it places us on Mars. We conceive this place as a "trampling around" activity–setting. These "boot prints" mark this place behind the Columbia Hills, tens of millions of kilometers away, as a place where people have been. The photograph symbolizes what we feel in exploring Mars through the rovers; in the words of MER scientist, Jim Rice:

Fig. 13 "The Footprint on Mars" – wheel tracks of the Spirit Mars Exploration Rover (Credit: NASA/JPL)

> I put myself out there in the scene, the rover, with two boots on the ground, trying to figure out where to go and what to do, and how to make that what we're observing with the instruments. Day in day out, it was always the perspective of being on the surface and trying to draw on your own field experience in places that might be similar. (Clancey 2012, p. 100)

Conceptualization is inherently value-laden and hence self-referential: What does this place, thing, event, idea, etc. mean to me? We encounter a place in a manner that fits our ways of knowing within a purposeful activity and conceive it accordingly. Scientists structure their inquiry, particularly in observing, recording, and organizing data, so that it may be accommodated by established schemas of scientific models and social order. For example, data collection during the Mars Phoenix mission was adjusted to create tables and comparative analyses urged by scientific journal editors. Consequently, Mars as a place was seen-through the schemas and constraints imposed by the activity of scientific publication. What was interesting to photograph, shovel, sample, and measure near the Phoenix lander became ordered by the layout of tables and graphs. My own photographs from Haughton-Mars Expeditions and FMARS were sometimes taken because of a striking perspective or lighting, but also generally just attempting to "map" with images where people were and what they were doing.

More generally, how graphics are used during these missions reveals different ways that people can relate to Mars and the scientific work (Fig. 14). We can

Fig. 14 Visual artifacts reveal the different dimensions by which we conceive of a place, creating a setting – what are we seeing the place as, what activity does it afford? (Credits: NASA/JPL and Jim Bell)

categorize graphics in terms of purposes and activities – what are these people doing or expressing? We find a range of motives from "pure" artistic expression to scientific inquiry–but *personal feeling and aesthetics (what looks good) are important for all purposes*. These images show how Mars as a place can be conceived *practically* for depicting or planning a route (top-left photograph: creating a synthetic, third-person view by superimposing the rover in a photograph taken further down the hill; also bottom-right: superimposing a route on a photograph taken from orbit); *scientifically*, by analyzing chemical structure (top center); *artistically*, by deliberately creating and selecting pleasing photographs (notably Bell's coffee-table book and the commanding view from the Columbia Hills).

In summary, aesthetically pleasing images (interesting, intriguing, evocative, and/or awe-inspiring) have these qualities because *they convey information through feelings coupled with multiple categorical relations* – including relating perception and action (e.g., rover tracks), relative scale (the two left-most photographs), color (particularly the choice of artificial color in the spectral analysis), and 3-D perspective (the orbital view). The emotional quality of being pleasing accompanies a non-discursive, meaningful way of knowing a place. The initial experience is feeling value (Damasio 1994, 1999), which tacitly primes attention to study the image. Looking at the spectral analysis or the view from the Columbia Hills, we experience its significance and may express our appreciation of its value. Situated cognition theory suggests that what constitutes information, what is significant, arises within an ongoing, mentally organized conception of activities and our concomitant identities. In short, who we are, what we are doing now, what interests us, how we see the world, how we feel about it and what we claim to perceive, how we interpret our perceptions, and the possible actions are all coupled, mutually dependent and emerging as a whole within our mentality. Our sequential actions of looking, describing, manipulating – our inquiry – consequentially enables further ways of knowing and interpreting the place, which we will represent by additional images, descriptions, models, and artistry.

5 Historical Precedents and Theories Relating Setting, Identity, and Felt Experience

Socio-technical studies of work settings have been informed by earlier research in proxemics and urban design; both provide data, methods, and application opportunities for situated cognition analyses. Past and present studies have a common methodological orientation, termed "design for activities" or "design for interaction," which considers physical needs in the environment (shelter, shade, seating, etc.) and the practical needs of social interactions (e.g., conversation, chance meeting, eating together). *Proxemics* is defined broadly in the context of animal behavior (ethology) as the study of nonverbal communication, such how distance and orientation when standing near someone reflects a cultural attitude of intimacy (or not).

In *urban design* sociologists and architects incorporated humanist, activity-oriented perspectives in public settings, such as rail station waiting rooms (Hall 1966) and city plazas (Whyte 1980) to encourage improvised sitting and clustering. Hall characterized these "reaction bubbles" as "interrelated observations and theories of man's use of space as a specialized elaboration of culture" (p. 129, p. 1). Hall believed that the value in studying proxemics is that it enables evaluating not only the way people interact with others in daily life, but guides design of "the organization of space in [their] houses and buildings, and ultimately the layout of [their] towns." This claim is reflected by Lynch's (1955, 1960/2001) study of Boston, in which he promoted.

> [...] consideration of the city as a complete landscape that is seen, felt, and heard as a complicated sensuous environment that encompasses us throughout much of our life. What is the effect on us of all that we sense while we loiter or bustle through the city streets and squares? What can we do to make this flow of stimuli more satisfying, more inspiring, more humane? (Lynch 1955, archival page "K.L. 3-8-55 1955 March 8"; Valdez 2011).

In sociotechnical studies of the 1990s, anthropologists – who had realized long ago that layout, settings, and interactions with a physical environment were integral to behavior, ritual, and culture more generally – recorded and described how activities in workplaces were socially ordered (e.g., Jordan and Henderson 1995), interpreting proxemics as manifesting relationships (e.g., a feeling of discord, parsimony, or threat).

Alexander et al.'s (1977) notion of a design "pattern language," which has been influential in a wide range of disciplines including computer programming, is particularly relevant in understanding settings. He emphasizes that patterns in successful urban designs and homes are organic, they emerge naturally through and during activities. We can design objects, architecture, and processes, but uses, values, adaptations, and practices will develop through experience over time. As we see in the HMP base camp, the structure and functions of settings, like the behaviors that occur there, are dynamic. Meanings and functions are reinforced through interaction but also prone to improvised or reflective modification as new perceptions and conceptions develop, often prompted or constrained by changing contexts (physical and social). A simple example in the FMARS habitat is how the upper deck table has been positioned. Originally it was oriented lengthwise, perpendicular to the wall in front of the large portal; by 2001 it was parallel to the wall, allowing one to turn around from the table to the workbench behind.

Alexander's theory provides a way of understanding aesthetic, pleasing designs as having multiple levels of patterned organization:

> In short, no pattern is an isolated entity. Each pattern can exist in the world, only to the extent that it is supported by other patterns: the larger patterns in which it is embedded, the patterns of the same size that surround it, and the smaller patterns which are embedded in it. This is a fundamental view of the world. It says that when you build a thing you cannot merely build that thing in isolation, but must also repair the world around it, and within it, so that the larger world at that one place becomes more coherent, and more whole; and the thing which you make takes its place in the web of nature as you make it. (Alexander et al. 1977, p. xiii)

For example, Wurster's pattern language for home architecture (Treib 1995/1999) included a backbone serving as a corridor connecting private spaces with public living and dining rooms in between, typically created by expanding the corridor towards the back of the house. Further, rooms were designed and oriented to the environment such that the outdoors became an extension of inside; it is experienced and lived in as one setting with nested parts.

A situated cognition perspective relates well to this idea of "deeply rooted patterns" in what Alexander calls "timeless ways of building." Relationships that Alexander finds in design of alcoves and roads relate perception, emotion, and activity in a manner that is also inherently social (e.g., a setting designed for privacy, itself a social relation, may also enable "watching life"). Patterns are not rationally inferred or necessarily articulated, but embodied in how we relate to and configure furniture with respect to windows in a room or arcades in a garden. To demonstrate this duality, in the quoted excerpt above we can substitute "activity" or "practice" for pattern, "automated tool" for thing, "setting" for world, and "human experience" for nature:

> In short, no activity is an isolated entity. Each activity can exist in a setting, only to the extent that it is supported by other activities: the larger practices in which it is embedded, the activities of a similar nature that surround it, and the smaller actions which are embedded in it. This is a fundamental view of the world. It says that when you build a tool you cannot merely build that tool in isolation, but must also repair the setting and practices around it, and within it, so that the larger world at that one place becomes more coherent, and more whole; and the tool which you make takes its place in the web of human experience as you make it.

This substitution is another way of showing that settings, activities, and identities are conceptually coupled: The visible physical *patterning of things* we create and configure (e.g., the base camp, the work tent) and the *patterning of our behavior* in the world (e.g., practices in the work tent; sitting on the ATVs) both reflect and shape the *patterning in our understanding* of the social world as perceptual and linguistic categories and conceptions of settings, activities, roles and values. This dynamic is summarized well by social scientists engaged in "design ethnography":

> The main virtue of ethnography is its ability to make visible the "real world" sociality of a setting. As a mode of social research it is concerned to produce detailed descriptions of the "workaday" activities of social actors within specific contexts. It is a naturalistic method relying upon material drawn from the first-hand experience of a fieldworker in some setting. It seeks to present a portrait of life as seen and understood by those who live and work within the domain concerned. It is this objective which is the rationale behind the method's insistence on the direct involvement of the researcher in the setting under investigation. The intention of ethnography is to see activities as social actions embedded within a socially organised domain and accomplished in and through the day-to-day activities of participants. (Hughes et al. 1994, p. 430; quoted by Rönkkö 2010)

6 Conclusions About Places, Situations, and Activities

In this section I reflect on the editors' original topic summary for this book, explaining how I interpret the various themes and questions.

Socio-technical studies are typically interdisciplinary collaborations that "bring together contributions on the *spatiotemporal contingency of human life* from different fields of research."[4] Notably in today's work system design teams, a cognitive psychologist might be carving up practices according to "tasks"; a human factors psychologist emphasizing "workload" and fatigue; a social anthropologist categorizing settings, activities, impasses, and circumstantial interactions; a sociologist analyzing conversational patterns; and a computer scientist mapping workflow, data structures, and communication media. In this manner, the practical, functional, epistemological, emotional, social, and physical "facets" and motifs of life are reified in descriptions and designs.

To recapitulate basic terminology: A *situation* for an actor is conceptual, a mental construct. In contrast, in socio-technical studies, *places* are names given by a community of practice to a physical environment (both natural and built) as a means of structuring their knowledge and activities. Examples include Haughton Crater, "the lower deck of the hab," and "the work tent." Places in a cultural parlance provide a shared frame of reference for describing the world and activities; consequently, many cultures may share the same distinctions, such as the map of the world (though disagreeing about some boundaries). *Settings* for actors in a culture are conceptually both more abstract (e.g., "a stage," "a retail store," "an expedition camp") and elaborated insofar as they always entail certain activities and not others. Thus in our understanding, our "mental models" of the world and our practices, settings are conceptually tied to norms and identities.

In practice, a particular activity may always occur in a particular place, such that culturally the name of the place, the activity, and its meaning as a setting blend. For example, for over 15 years the Haughton Mars Project organization set up a permanent camp with buildings and rigid tents, such that "the HMP base camp" designated a specific place on Devon Island. This prescribed area became the setting for "the HMP expedition," an annual summer activity.

Illustrating that places can be known universally but settings are cultural distinctions, the HMP was required by the Inuit to move from its original river site to an area outside the crater in 2000. The Nunavut Territory owns this land and the Inuit people made known that Haughton Crater was sacred. The boundaries of the place marked on the map were universally agreed, but the native and scientist communities had completely different understandings of its significance as a setting for activities. The Nunavut council on neighboring Ellesmere Island decided that the HMP expedition and Mars Society would have only limited access to the crater. Thus the crater as a setting for activities is a direct expression of identity, in which the scientific and native "tribes" seek to express "who we are" and "what we do" in the same

[4] From "Rationale – Situatedness and Place." Cf. also the *Preface* of this volume.

place, construing it as a setting in very different ways. Strikingly both social groups feel the "spirituality" of the place and desire to leave it undeveloped and pristine, which provided common ground for a compromise. (A similar story is playing out today with astronomers on Mauna Kea volcano in Hawaii.)

Of course, cultures regulate their own activities within settings, too. HMP camp was structured like a tiny village so different activities occurred in their own setting – the mess tent for cooking and eating, the greenhouse, the location for placing personal tents, etc. However, the community may also view these as places where different activities are negotiable and open, such that the meaning of the place as a setting is adapted and interpreted over time through practice (e.g., the mess tent is reconceived and reconfigured as a setting for after-dinner lectures; the Inuit allowed scientific work within Haughton Crater if the scientist is accompanied by a member of their community).

In summary, actors' knowledge of "what happens in a place" is essentially how they know the place as a setting, for being "who we are." In some respect activities can be easily articulated (e.g., "expedition members sit and work at their laptop computers in the work tent"); in other respects, purposes may be tacit, not reflected upon, or individually improvised (e.g., you can use your work area to store personal things you need during the day, such as your tooth brush and a towel).

I have also stressed that in my interpretation (Clancey 1997) *situatedness* refers to the capability to conceptually organize behavior that relates feelings, perception, and understanding of the self as a social actor: What am I doing now? (WIDN) Who am I being now? (WIBN) How should I be behaving now? How well am I doing now? This ongoing, multidimensional categorizing capability, falling under the rubric *higher-order consciousness* (see Clancey 1993), is a neuropsychological process (its mechanism) and pervasively social in orientation and meaning (its content). Thus the "situatedness of human life" is not a claim about a *place*, because "being situated" is an ongoing conceptual-interactive accomplishment, a coordination of understanding and behavior dynamically, in real time. Part of this *conceptual coordination* involves continuously reconceiving "the situation" to simultaneously fit and construct one's identity, in all the respects listed above – WIDN, WIBN, etc.

In summary, my understanding of "the situation" (or more completely, "my situation as a social actor, here, now") means much more than being in a physical place, which of course existence trivially entails. Similarly, saying cognition is *embodied*, is not the trivial claim of "being in a body," but that activity is conceptually coordinated with respect to perceptual-motor experience, manifest particularly in body-oriented metaphors (Lakoff and Johnson 1980). Putting this together, the MER scientists are able to do field science because they are conceptually embodied in the rover, and they coordinate the rover's actions through a broad and particular understanding of the field science activity through which they conceive "the situation we are in now" – all without being in the place where the investigation is occurring, on Mars.

Thus, my *understanding of the setting* in which I carry out my activity (here, now) transcends my objective designation of the place. The MER scientists conceive of Meridiani Planum and Gusev Crater, named places appearing on maps of

Mars, as complicated, finely structured settings (e.g., an ancient streambed; outcroppings marking a geological disconformity) for applying their roving laboratory instruments, in the activity of surveying and probing the story of water and possible life on Mars. They know the place and hence who they are ("what I am doing now") by how they can move through it and analyze its chemical and atomic nature. Thus places on Mars where they work, are conceptually transformed by their images and tentative interpretations into settings for field science. These practices and competence makes them Mars explorers.

Simultaneously, for the scientists working in person together in Pasadena, California during the first months of each expedition, being at JPL in the privileged science meeting rooms (restricted to members of the team), conferred and reinforced an identity of participating on an historic mission, the first overland expedition on another planet. Even visiting the workplace, building 264, is awe inspiring, for it has been the setting for the Viking missions (the first landings on Mars), the Voyager missions (the first spacecraft to fly by the outer planets), the more recent New Horizon mission to Pluto, and many more planetary explorations. Thus, the meaning of the place, how and why it is valued, may develop such that it becomes sacred within the culture, which is to say the activities that occur there have an intrinsic, transcendent purpose for the community such that the name of the place evokes its social meaning as a hallowed setting.[5] JPL/264 has this meaning for planetary scientists. Symbolic analogs for Mars, like FMARS, where being on Mars can be imagined and practiced, have become pilgrimage settings for members of the Mars Society and other believers in the vision. The mystique of the planet Mars itself increasingly captures the public's imagination: it is aesthetically alluring, scientifically interesting, and a place where we can rove, explore, and someday might live. As we make Mars a setting for these activities, it makes us artists, scientists, and potential settlers. We conceive of the place in these ways and sometimes all at once, creating and realizing new identities for ourselves and the people who follow.

References

Alexander, C., S. Ishikawa, and M. Silverstein. 1977. *A Pattern Language: Towns, Buildings, Construction*. New York: Oxford University Press.

Anderson, R. 1994. Representations and requirements: The value of ethnography in system design. *Human-Computer Interaction* 9 (3): 151–182.

———. 1997. Work, ethnography and system design. In *The Encyclopedia of Microcomputers*, ed. A. Kent and J.G. Williams, vol. 20, 159–183. New York: Marcel Dekker.

Barker, R.G. 1968. *Ecological Psychology: Concepts and Methods for Studying the Environment of Human Behavior*. Stanford: Stanford University Press.

[5] For a related description of expeditions as liminal experiences see Clancey (2012, Chapter 9), particularly the references to Martin Rudwick's analysis.

Bartlett, F.C. 1932/1977. *Remembering – A Study in Experimental and Social Psychology*. Cambridge: Cambridge University Press.
Clancey, W.J. 1993. The Biology of Consciousness: Comparative Review of Israel Rosenfield, The Strange, Familiar, and Forgotten: An Anatomy of Consciousness and Gerald M. Edelman, Bright air, Brilliant Fire: On the Matter of the Mind. *Artificial Intelligence* 60: 313–356.
———. 1997. *Situated Cognition: On Human Knowledge and Computer Representations*. New York: Cambridge University Press.
———. 1999. *Conceptual Coordination: How the Mind Orders Experience in Time*. Mahwah: Erlbaum.
———. 2001. Field science ethnography: Methods for systematic observation on an arctic expedition. *Field Methods* 13 (3): 223–243.
———. 2002. Simulating activities: Relating motives, deliberation, and attentive coordination. *Cognitive Systems Research 3 (3)*: 471–499. special issue "Situated and embodied cognition".
———. 2006a. Observation of work practices in natural settings. In *Cambridge Handbook on Expertise and Expert Performance*, ed. A. Ericsson, N. Charness, P. Feltovich, and R. Hoffman, 127–145. New York: Cambridge University Press.
———. 2006b. Participant observation of a Mars surface habitat simulation. *Habitation: International Journal for Human Support Research* 11 (1/2): 27–47.
———. 2008. Scientific antecedents of situated cognition. In *Cambridge Handbook of Situated Cognition*, ed. P. Robbins and M. Aydede, 11–34. New York: Cambridge University Press.
———. 2012. *Working on Mars: Voyages of Scientific Discovery with the Mars Exploration Rovers*. Cambridge, MA: MIT Press.
———. 2014. Working on Mars: Translating planetary field science to our distant lands. *KERB Journal of Landscape Architecture 22*: 56–59. Special issue "Remoteness".
Clancey, W.J., P. Sachs, M. Sierhuis, and R. van Hoof. 1998. Brahms: Simulating practice for work systems design. *International Journal of Human-Computer Studies* 49: 831–865.
Clancey, W.J., M. Sierhuis, B. Damer, and B. Brodsky. 2005. Cognitive modeling of social behaviors. In *Cognition and Multi-agent Interaction: From Cognitive Modeling to Social Simulation*, ed. R. Sun, 151–184. New York: Cambridge University Press.
Clancey, W.J., C. Linde, C. Seah, and M. Shafto. 2013. *Work Practice Simulation of Complex Human-Automation Systems in Safety Critical Situations: The Brahms Generalized Überlingen Model.* NASA Technical Publication 2013-216508, Washington, DC.
Damasio, A. 1994. *Descartes' Error: Emotion, Reason, and the Human Brain*. New York: Putnam.
———. 1999. *The Feeling of What Happens: Body and Emotion in the Making of Consciousness*. New York: Harcourt Brace and Company.
De Botton, A. 2002. *The Art of Travel*. New York: Pantheon.
Dewey, R. 2014. Hacking remoteness through viewpoint and cognition. *KERB Journal of Landscape Architecture* 22: 26–33.
Downes, S. 1990. Herbert Simon's computational models of scientific discovery. In *Proceedings of the Biennial Meeting of the Philosophy of Science Association*, 97–108. Chicago: University of Chicago Press. Retrieved from http://www.jstor.org/stable/192696.
Fox, W.L. 2000. *The Void, the Grid, and the Sign: Traversing the Great Basin*. Salt Lake City: University of Utah Press.
Greenbaum, J., and M. Kyng, eds. 1991. *Design at Work: Cooperative Design of Computer Systems*. Hillsdale: Lawrence Erlbaum Associates.
Hall, E.T. 1966. *The Hidden Dimension*. New York: Doubleday.
Harper, R. 2000. The organisation in ethnography: A discussion of ethnographic fieldwork programs. *Computer Supported Cooperative Work* 9 (2): 239–264.
Hughes, J., T. King, T. Rodden, and H. Andersen. 1994. Moving out from the control room: Ethnography in systems design. In *Proceedings of computer supported work*, October 22–26, 417–428. New York: ACM.
Jordan, B., and A. Henderson. 1995. Interaction analysis: Foundations and practice. *Journal of the Learning Sciences* 4 (1): 39–103.

Keltner, D., and J. Haidt. 2003. Approaching awe, a moral, spiritual, and aesthetic emotion. *Cognition and Emotion* 17: 297–314.

Lakoff, G., and M. Johnson. 1980. *Metaphors We Live By*. Chicago: University of Chicago Press.

Langley, P.W., H.A. Simon, G. Bradshaw, and J.M. Zytkow. 1987. *Scientific Discovery: Computational Explorations of the Creative Process*. Cambridge: MIT Press.

Lave, J. 1988. *Cognition in Practice*. Cambridge: Cambridge University Press.

Lee, P. and G.R. Osinski. 2005. The Haughton-Mars Project: Overview of science investigations at the Haughton impact structure and surrounding terrains, and relevance to planetary studies. *Meteoritics & Planetary Science* 40 (12):1755–1758.

Leont'ev, A.N. 1979. The problem of activity in psychology. In *The Concept of Activity in Soviet Psychology*, ed. J.V. Wertsch, 37–71. Armonk: M.E. Sharpe.

Luff, P., J. Hindmarsh, and C. Heath. 2000. *Workplace Studies: Recovering Work Practice and Informing System Design*. Cambridge: Cambridge University Press.

Lynch, K. 1955. Perceptual form of the city. *Kevin Lynch Papers*, Massachusetts Institute of Technology, Institute Archives and Special Collections (Cambridge, Massachusetts, United States), MC 208, Box 1, General Notes.

———. 1960/2001. *The Image of the City.* Cambridge: MIT Press.

Mikulak, A. 2015. All about awe: Science explores show life's small marvels elevate cognition and emotion. *Observer* 28 (4): 16–19.

Nardi, B. 1996. *Context and Consciousness: Activity Theory and Human-Computer Interaction*. Cambridge, MA: MIT Press.

Rönkkö, K. 2010. Ethnography. In *Encyclopedia of Software Engineering*, ed. P. Laplante. New York: Taylor and Francis Group.

Ryle, G. 1949. *The Concept of Mind.* New York: Barnes & Noble.

Salvador, T., G. Bell, and K. Anderson. 1999. Design ethnography. *Design Management Journal* 10 (4): 35–41.

Spradley, J.P. 1980. *Participant Observation*. Fort Worth: Harcourt Brace College Publishers.

Treib, M. ed. 1995/1999. *An Everyday Modernism: The Houses of William Wurster.* Berkeley: University of California Press.

Valdez, R. 2011. *Seattle's Land Use Code: Kevin Lynch on Cities*. April 11, 2011. Retrieved from https://seattleslandusecode.wordpress.com/2011/04/25/kevin-lynch-on-cities/. Accessed 15 Nov 2016.

Wenger, E. 1998. *Communities of Practice: Learning, Meaning, and Identity*. New York: Cambridge University Press.

Wertsch, J.V., ed. 1979. *The Concept of Activity in Soviet Psychology*. Armonk: M.E. Sharpe.

Whyte, W.H. 1980. *Social Life of Small Urban Spaces*. Washington, DC: Conservation Foundation.

Zubrin, R. 2003. *Mars on Earth: The Adventures of Space Pioneers in the High Arctic*. New York: Tarcher Penguin.

The Place of Mind

Thomas Hünefeldt

The mind is its own place

John Milton: *Paradise Lost,* Book 1, 254

Abstract In both past and contemporary philosophy and science, the question as to where mind takes place has received an amazingly large variety of answers. In this chapter, I will first introduce some important distinctions concerning the variables that are logically involved in this question (1.), then sketch a brief and necessarily rough systematic overview of the basic kinds of answers that have been given to this question (2.), and finally attempt to give a phenomenological account of these basic kinds of answers (3.).

Keywords Mind · Mental states and processes · Consciousness · Place · Situatedness · World · Phenomenology · Cognitive science

Today it is a widely accepted and even commonplace view that mind takes place in the brain. In fact, mental states and processes are generally believed to emerge from, or to be constituted by, neurophysiological states and processes in the brain. Accordingly, individual minds are generally believed to emerge from, or to be constituted by, the neurophysiological states and processes in individual brains or, more generally, in sufficiently complex nervous systems. Though this view is nowadays often taken for granted, it is only during the second half of the twentieth century that it has become predominant and it is far from being uncontroversial even today. As a

T. Hünefeldt (✉)
ECONA – Interuniversity Center for Research on Cognitive Processing in Natural and Artificial Systems, Sapienza University of Rome, Rome, Italy

Interdisciplinary Research Group on "Philosophy of Place", Catholic University of Eichstätt-Ingolstadt, Eichstätt, Germany
e-mail: thomas.huenefeldt@uniroma1.it

T. Hünefeldt, A. Schlitte (eds.), *Situatedness and Place*, Contributions To Phenomenology 95, https://doi.org/10.1007/978-3-319-92937-8_7

matter of fact, both in past and in contemporary philosophy and science, the question as to *where* mind takes place has received an amazingly large variety of answers. While this question has traditionally often been answered in terms of different ways to localize the "seat" of (the) mind,[1] it has recently often been discussed in terms of different ways to conceive the "situatedness" of cognition and other mental states and processes, their "embodiment" in the brain or the body, and their potential "extension" beyond the brain and the body.[2] In this chapter, I will first introduce some important distinctions concerning the variables that are logically involved in the question as to *where* mind takes place (1.), and then sketch a brief and necessarily rough systematic overview of the basic kinds of answers that have been given to this question (2.). Rather than in arguing for or against particular kinds of answers, I am interested in exploring why different kinds of answers are possible at all and how different kinds of answers are related despite the radical differences between them (3.).

1 Conceptual Framework

From a purely logical point of view, the question as to *where* mind takes place may have basically three different kinds of answers: *everywhere*, *nowhere*, or only *somewhere* but not everywhere. Before exploring these three basic kinds of potential answers, it will be helpful to introduce some important distinctions concerning three variables that are logically involved in this question: (1) the concept of *mind* that is being supposed, (2) the concept of the all-encompassing place of a *world* wherein mind may be supposed to take place either everywhere, nowhere, or somewhere but not everywhere,[3] and (3) the concept of *taking place* therein.

1.1 The Concept of Mind

In the question as to where mind takes place, the term "mind" is meant to refer to an indefinite number and kind of mental states and processes. The concept of mind that is being supposed in this question is thus characteristically different from, but not

[1] Much of modern reference to a "seat" of (the) mind has been influenced by Descartes who notably referred to a "seat of the soul" (*siège de l'âme*), locating it in the "pineal gland" of the human brain (cf., e.g., Descartes (1649), *The Passions of the Soul*, Art. 32ff). Ancient authors such as Aristotle preferred referring to the "place" (*tópos*) wherein particular kinds of mental states and processes (e.g., "sensations") have their "origin" or governing "principle" (*arché*) (cf., e.g., Aristotle (1961), *Parts of Animals*, II, 656a). Cf. also Lind (2007) and Onions (1951) for reviews of ancient views on the "seat" of mind.

[2] Cf. Adams and Aizawa (2010), Barrett (2011), Clark (2008), Gallagher (2005), Menary (2010), Mesquita et al. (2011), Robbins and Aydede (2009), Rupert (2009), and Shapiro (2011, 2014), for monographies on these issues.

[3] Cf. the discussion on the relationship between place and world in the Introduction to this book.

incompatible with, the concept of mind that is implied in reference to *a* mind, i.e. to one or more minds.[4] In the latter context, in fact, the term "mind" refers to some kind of whole involving a plurality of mental states and processes (e.g., to the whole of the mental states and processes, or of the mental faculties, of a particular individual). Now, depending on how mental states and processes are conceived, both the former and the latter concept of mind may be either a common-sense or a scientific concept of mind.

According to the common-sense concept of mind, mental states and processes are states and processes such as feeling, desiring, willing, perceiving, remembering, imagining, believing, reasoning, planning, etc. Such states and processes have in common that they may be considered either from a subjective (or first-person) perspective or from an objective (or third-person) perspective. From a subjective (or first-person) perspective, mental states and processes appear as subjectively lived states and processes, i.e. as states and processes that constitute particular aspects of one's own world. For example, seeing appears as a state or process which constitutes the visual aspect of one's own world, whereas hearing appears as a state or process which constitutes the acoustic aspect of one's own world, etc. From an objective (or third-person) perspective, by contrast, mental states and processes appear as particular kinds of objectively manifest states and processes, i.e. as particular kinds of states and processes concerning particular kinds of inner-world entities which influence the response of these entities to particular kinds of stimuli under particular kinds of circumstances. For example, seeing appears as a state or process concerning particular kinds of inner-world entities (e.g., human beings) which renders possible their response to visual stimuli, whereas hearing appears as a state or process concerning particular kinds of inner-world entities (e.g., human beings) which renders possible their response to acoustic stimuli, etc.

In addition to distinguishing between a subjective (or first-person) and an objective (or third-person) perspective on mental states and processes, it is common practice to distinguish between conscious and unconscious mental states and processes, and to refer to this distinction both from a subjective (or first-person) and from an objective (or third-person) perspective. From a subjective (or first-person) perspective, unconscious mental states and processes are necessarily related to conscious mental states and processes, in so far as subjectively living something amounts to having consciousness of it. From an objective (or third-person) perspective, by contrast, mental states and processes are not necessarily related to consciousness, in so far as inner-world entities which manifest behavior typically associated with particular kinds of mental states and processes (e.g., seeing or hearing) need not be conscious, i.e. they need not be subjectively living these or any other mental states and processes. Accordingly, consciousness is usually supposed to be a particular kind of mental states and processes, which renders possible subjectively living something and having a subjective world.

[4] In ordinary language, the noun "mind" is often accompanied by an article (e.g., *a* mind, *the* mind, etc.), by a possessive adjective (e.g., *my* mind, *your* mind, *her* mind, etc.), or by a demonstrative (e.g., *this* mind, *that* mind, etc.), and it can thus be used either in singular or in plural (e.g., *one* mind, *two* minds, etc.).

In contrast to this common-sense view of mental states and processes, in general, and consciousness, in particular, scientific views tend to conceive mental states and processes, in general, and consciousness, in particular, exclusively from an objective (or third-person) perspective and to refer to a subjective (or first-person) perspective, if at all, only in so far as it is objectively manifest, for example in verbal accounts. Furthermore, the objectively manifest states and processes that common-sense views, on the one hand, and scientific views, on the other, refer to as mental states and processes need not be identical, but may be – and in fact often are – more or less different. For example, common-sense views generally refer to particular kinds of states and processes displayed by human beings and other kinds of animals in natural settings, whereas scientific views generally refer either to the more or less different kinds of states and processes displayed by various kinds of natural or artificial systems in experimental settings, or to the completely different kinds of states and processes displayed by particular subsystems of such systems, e.g. to neurophysiological or computational states and processes. Consequently, the common-sense concept and the scientific concept of mental states and processes generally overlap only partly.

1.2 The Concept of World

The concept of *world* that is implied in the question as to where mind takes place refers to a world wherein one's own and other subjects' mental states and processes may be supposed to coexist as states and processes concerning particular inner-world entities. It therefore refers to some kind of objective world, i.e. to the world that is actually given and lived as it is represented to be independently of any particular subject. In other words, due to its reference to the concept of mind, the question as to where mind takes place is not a phenomenological but an ontological question and thus requires an ontological perspective on the world that is actually given and lived.

Now, from a fundamental ontological perspective, all reality may be supposed to be ultimately constituted either by a single, all-encompassing substance or by a plurality of substances, thus giving rise to either monistic or pluralistic ontologies. Here and throughout this chapter, the term "substance" is used in its broad, etymological sense which refers to whatever is supposed to "stand under or ground things".[5] It does therefore not necessarily refer to some fundamental object or thing, but it may as well refer to some fundamental property, event, or process. In fact, it here refers to whatever is supposed to be a basic constituent of reality. Accordingly, the terms "monistic" and "pluralistic" ontology here refer to either existence or priority monism and pluralism, respectively. Whereas "existence monism" may be defined as "the doctrine that exactly one concrete object token exists", "priority

[5] Cf. Robinson's (2014) entry on "substance" in the *Stanford Encyclopedia of Philosophy*.

monism" may be defined as "the doctrine that exactly one concrete object is basic".[6] I here adapt these definitions by referring to the concept of "substance" in its broader, etymological sense rather than to the concept of "concrete object".

Furthermore, supposing the distinction between physical and mental predicates, a substance may be conceived either in purely physical terms, or in purely mental terms, or in both physical and mental terms, as a substance that has both physical and mental properties or aspects, thus allowing to distinguish between different versions of either monistic or pluralistic ontologies. In particular, these different conceptions allow distinguishing not only between materialist, idealist, property dualist, and dual aspect neutral monist versions of either monistic or pluralistic ontologies, but also between substance monist and substance pluralist versions of pluralistic ontologies, depending on whether the supposed plurality of substances are supposed to fall under one or more highest types.[7] The substance monist versions include the materialist, idealist, property dualist, and dual aspect neutral monist versions of pluralistic ontologies, whereas the substance pluralist versions of pluralistic ontologies include, among others, ontologies generally known as "substance dualism", i.e. ontologies supposing that there are two highest types of basic substances: the physical and the mental.

To give some example, Lemaître's (1927) "big bang theory" in physical cosmology, Berkeley's (1710) "subjective idealism", Spinoza's (1677) pantheism, and Hegel's (1817) "absolute idealism" might arguably be considered as materialist, idealist, property dualist, and dual aspect neutral monist versions of monistic ontologies, respectively, whereas the "steady-state theory" (Bondi and Gold 1948, Hoyle 1948) in physical cosmology, Leibniz's (1714) "monadology", Lotzes' (1852, 1856) atomistic panpsychism, and Mach's (1886) neutral monist constructivism might arguably be considered as materialist, idealist, property dualist, and dual aspect neutral monist versions of pluralistic ontologies, respectively. Furthermore, the latter four examples might be considered as different kinds of substance monist versions of pluralistic ontologies, whereas Descartes' (1649) substance dualism might be considered as a particular kind of substance pluralist (namely a substance dualist) version of pluralistic ontologies, etc.

1.3 The Concept of Taking Place

Finally, the concept of *taking place* refers to some way of being somewhere. Evidently, there may be many different ways of being somewhere, depending on *what* is supposed to be there and on *where* it is supposed to be, i.e. depending on the ontology that is supposed. In my own world, for example, feelings take place

[6] Cf. Schaffer's (2016) entry on "monism" in the *Stanford Encyclopedia of Philosophy*.

[7] As opposed to "existence monism" and "priority monism", "substance monism" may be defined as "the doctrine that all concrete objects fall under one highest type (perhaps material, or mental, or some neutral underlying type […])" (Cf. Schaffer 2016).

differently than things, stones take place differently than raindrops, and lightning takes places differently than thunder, and I suppose that the same is true in your world and in the world of many other subjects. Analogously, there may be many different ways of taking place in the objective world, depending on how that world is conceived. For example, each of the four elements proposed by Empedocles ("fire", "earth", "air", and "water") takes place differently from each other, Democritus's "atoms" take place differently than the "quarks" supposed in contemporary physics, and material substances such as "atoms" or "quarks" take place differently than immaterial substances such as Leibniz's (1714) "monads" or Descartes' (1649) "souls".

In addition to such differences depending on the ontology that is supposed, there is at least one difference that applies across different kinds of ontologies to any conceivable world and that is relevant to our question as to where mind takes place. In fact, taking place may be conceived either as taking place in *some particular place* in a world or as taking place in *the all-encompassing place* of that world as a whole. For example, particular things and events such as physical atoms or chemical reactions are supposed to take place in particular places in the objective world, whereas the conservation of energy, the increase of entropy, or the metric expansion of space are supposed to take place in the all-encompassing place of the objective world as a whole. Accordingly, it is possible to distinguish between local states and processes, which take place in in particular places in a world, on the one hand, and global states and processes, which take place in the all-encompassing place of that world as a whole, on the other.

Now, supposing the distinction between physical and mental predicates, both local states and processes and global states and processes may be conceived to take place either in terms of physical states and processes or in terms of mental states and processes. For example, local states and processes (e.g., the behavior of animate or inanimate things) may be conceived to take not only in terms of physical states and processes such as chemical reactions or neuronal discharges, but they may as well be – and have in fact often been – conceived in terms of mental states and processes such as human loving, thinking, or willing. Analogously, global states and processes may be conceived to take place not only in terms of physical states and processes such as the conservation of energy, the increase of entropy, or the metric expansion of space, but they may as well be – and have in fact often been – conceived in terms of mental states and processes such as a divine love, a purposeful rational world-mind, or an anonymous irrational world-will.

2 A Systematic Overview

As mentioned above, the question as to *where* mind takes place may have, from a purely logical point of view, basically three different kinds of answers: *everywhere*, *nowhere*, or *somewhere* but not everywhere. In the following, I will explore in how far these three basic kinds of answers are matched by the answers that have in fact been given to this question in the history of philosophy and science, in general, and in contemporary philosophy and science, in particular. In searching for such "matches", it will of course neither be possible to attain completeness nor to do justice to alternative interpretations. What I am interested in is in fact rather to illustrate and to further articulate the spectrum of possible answers to the question as to *where* mind takes place.

2.1 Everywhere

The view that mind takes place *everywhere* in the objective world is generally known as "panpsychism", which may be defined as "the doctrine that *mind*, in some sense of the term, is *everywhere*, in some sense of that term".[8] Thus broadly conceived, panpsychism may be associated with two radically different kinds of ontologies which may be roughly illustrated by the two elliptical expressions "mind is in everything" and "everything is in mind". More precisely, it may be associated with either idealist or property dualist versions of either pluralistic or monistic ontologies.[9] On the one hand, in fact, panpsychism may be associated with the pluralistic thesis that all reality is ultimately constituted by a multitude of substances which are conceived either in purely mental terms as "units of experiencing" such as Leibniz's (1714) "monads", or in both physical and mental terms as substances that, as Lotzes' (1852, 1856) "atoms" or Mach's (1886) "elements", have both physical and mental properties or aspects. On the other hand, panpsychism may also be associated with the monistic thesis that all reality is ultimately constituted by an all-encompassing substance which is conceived either in purely mental terms as an idealist "world-mind" such as Berkeley's (1710) "God", or in both physical and mental terms as a substance that, as Spinoza's (1677) "God or Nature" or Hegel's (1817) "world-spirit", has both physical and mental properties or aspects.

These different kinds of panpsychism broadly conceived do not only imply fundamentally different ways of conceiving the mental states and processes that are supposed to take place everywhere, but also different ways of conceiving the "every-

[8] Cf. Seager and Allen-Hermanson's (2015) entry on "Panpsychism" in the *Stanford Encyclopedia of Philosophy*. For a recent comprehensive overview on panpsychism, see Seager (2018).

[9] Panpsychism is not always conceived thus broadly. According to Hartshorne (1950), for example, panpsychism "is the doctrine that everything is psychic or, at least, has a psychic aspect". It thus not only "contrasts with the monistic tendency of much idealism", but is also, "in its more significant form", not a "two-aspect theory", but "the view that all things, in all their aspects, consist exclusively of 'souls'". Similarly, Sprigge (1998) defines panpsychism as "the thesis that physical nature is composed of individuals each of which is to some degree sentient". In yet another sense, even Seager and Allen-Hermanson (2015), who provided the definition that we supposed above, give a somewhat narrower definition when, at the beginning of the same entry, they define panpsychism as "the doctrine that mind is a fundamental feature of the world which exists throughout the universe". In fact, the broader (and more literal) definition that we supposed above is not committed to the thesis that mind is "a fundamental feature" of reality, but admits the possibility that mind has a derived character, as long as the *ubiquity* of mind is granted.

where" where these mental states and processes are supposed to take place. In pluralistic kinds of panpsychism, the mental states and processes that are supposed to take place everywhere are states and processes concerning the multitude of substances that are supposed to constitute reality, so that "everywhere" really means "in every *thing*", but not necessarily "in every *place*". In fact, especially "atomistic" versions of panpsychism often admit some kind of "void space" "in between" the plurality of psychophysical "atoms" that they suppose to constitute reality. In monistic kinds of panpsychism, by contrast, the mental states and processes that are supposed to take place everywhere are states and processes concerning the all-encompassing substance that is supposed to constitute reality, so that the "everywhere" does not only mean "in every *thing*", but also "in every *place*". In fact, in monistic kinds of panpsychism, the all-encompassing substance is, as it were, the medium or place in which everything takes place.

Different kinds of panpsychism also differ in the kinds of mental states and processes that are supposed to take place everywhere. In pluralistic kinds of panpsychism, the mental states and processes that are supposed to take place everywhere are states and processes which concern each and every one of the substances that are supposed to constitute reality, so that they generally tend to be rather "basic" or "lower" mental states and processes such as sensation or sentience. In monistic kinds of panpsychism, by contrast, the mental states and processes that are supposed to take place everywhere are states and processes which concern the all-encompassing substance that is supposed to constitute reality, so that they generally tend to be rather "higher" mental states and processes such as thinking or willing. Furthermore, in pluralistic kinds of panpsychism, the substances that are supposed to constitute reality are generally supposed to be to some degree conscious, so that the mental states and processes that are supposed to take place everywhere generally include some at least basic kind of consciousness, and this consciousness is generally conceived from a subjective (or first-person) perspective according to which "[a]ny individual such that there is a truth *as to what it is like to be it* (in general or at some particular moment) is conscious, and its consciousness is what that truth concerns".[10] Accordingly, in pluralistic kinds of panpsychism, the mental states and processes that are supposed to take place everywhere are generally conceived rather in terms of subjectively lived states and processes than in terms of objectively manifest states and processes. In monistic kinds of panpsychism, by contrast, the mental states and processes that are supposed to take place everywhere are generally conceived rather in terms of objectively manifest states and processes than in terms of subjectively lived states processes and are in fact often conceived as unconscious states and processes such as a blind will or thought.

Both pluralistic and monistic kinds of panpsychism have to face the problem how to account for the common-sense distinctions between different individual minds, between different kinds of individual minds, and between entities with and without mental properties. More precisely, both pluralistic and monistic kinds of panpsychism have to explain whether, why, and how we should distinguish between

[10] Cf. Sprigge's (1998) account of panpsychism.

different individual minds, between different kinds of individual minds, and between entities with and without mental properties, and how these distinctions relate to the corresponding common-sense distinctions. Depending on the particular version of panpsychism, a large variety of answers have been given to these questions, and it is neither possible nor necessary to discuss them here. What is important here to note are two circumstances: first, most versions of panpsychism do indeed provide some answer to these questions, so that the view that mental states and processes take place everywhere is not only compatible with, but generally also explicitly involves the view that some kinds of entities (e.g., heaps of sand) have no mind of their own, and that some kinds of mental states and processes (e.g., seeing) take place only somewhere, but not everywhere; second, all answers to these questions imply an at least implicit reference to particular kinds of objectively manifest states and processes which allow to identify and individuate different individual minds and different kinds of individual minds.

Panpsychism, in any of its many guises, might seem a rather extravagant view today, but it is very ancient doctrine with a long and rich history in both western and eastern philosophy.[11] In western philosophy it was quite widespread throughout the nineteenth and well into the twentieth century. Indeed, panpsychist positions have been maintained by some of the most distinguished thinkers of these times and, significantly, in particular by some of the founders of scientific psychology such as Johann Friedrich Herbart (1828/1829), Rudolf Hermann Lotze (1852, 1856), Gustav Fechner (1851, 1861, 1946), Wilhelm Wundt (1863), and William James (1909, 1911). In recent times, panpsychism has found renewed interest in the context of attempts to tackle the "hard problem of consciousness" (Chalmers 1996) and it has been explicitly defended by a small but growing number of thinkers including Galen Strawson (2006a, b), David R. Griffin (1998), Gregg Rosenberg (2005), David Skrbina (2005) and Timothy Sprigge (1983, 2007).[12]

2.2 *Nowhere*

The view that mind takes place *nowhere* in the objective world amounts to the view that mental states and processes do not really exist. Though this view has never been espoused unconditionally, it may be associated with a radical version of the ontological position called "eliminative materialism", if the concept of "mind" is understood in a particular way. In fact, eliminative materialism may be defined as "the radical claim that our ordinary, common-sense understanding of the mind is deeply wrong and that some or all of the mental states posited by common-sense do not

[11] Cf. Seager and Allen-Hermanson (2015) and especially Clark (2004), Skrbina (2005), and Seager (2018) for overviews on the history of panpsychism.

[12] Cf. Seager and Allen-Hermanson (2015) and especially Skrbina (2009), Brüntrup and Jaskolla (2016) and Seager (2018) for overviews on contemporary panpsychism.

actually exist".[13] For eliminative materialists, the mental states and processes that do not really exist are thus the mental states and processes posited by common-sense, i.e. mental states and processes such as feelings, desires, intentions, thoughts, beliefs, etc. Similar to theoretical entities such as phlogiston or the luminiferous aether, such mental states and processes are supposed not to exist because they are supposed to be grounded in a radically false and outdated theory of mind.

Eliminative materialists would not deny that particular kinds of objectively manifest states and processes may be called *mental* states and processes, and that these kinds of objectively manifest states and processes may be more or less similar to the objectively manifest states and processes associated with the mental states and processes posited by common-sense. But for eliminative materialists these kinds of objectively manifest states and processes are all there is to mental states and processes, whereas any reference to subjectively lived states and processes would be at best a convenient fiction. In other words, eliminative materialists deny the existence of the mental states and processes posited by common-sense especially insofar as these states and processes are conceived not only in terms of objectively manifest states and processes but also in terms of subjectively lived states and processes.

On the one hand, eliminative materialism thus implies the view that mental states and processes conceived in terms of both objectively manifest and subjectively lived states and processes take place *nowhere* in the objective world. On the other hand, however, it also implies the view that mental states and processes conceived exclusively in terms of particular kinds of objectively manifest states and processes take place *somewhere* but not everywhere in the objective world. Accordingly, the view that mind, in *some* sense of the term, takes place *nowhere* in the objective world is not only compatible with, but generally involves the view that mind, in some *other* sense of the term, takes place *somewhere* but not everywhere in the objective world.

From an historical point of view, eliminative materialism seems to be "a relatively new theory with a very short history", which has been promoted in particular by contemporary "neurophilosophers" such as Paul and Patricia Churchland (1981, 1986, 1988).[14] Though it has received increasing attention with the rise of cognitive neuroscience in the last decades of the twentieth century, it has remained, at least so far, rather a minority view.

2.3 Somewhere

Finally, the view that mind takes place *somewhere*, but not everywhere, may be associated with virtually any kind of ontology. On the one hand, in fact, even those kinds of ontology which imply that mind, in some sense of the term, takes place either everywhere or nowhere generally also suppose that mind, in some *other* sense

[13] Cf. Ramsey's (2016) definition of "eliminative materialism" in the *Stanford Encyclopedia of Philosophy*.

[14] Cf. Ramsey (2016).

of the term, takes place somewhere but not everywhere. On the other hand, this view is more or less directly implied in virtually all other kinds of ontologies. Specifically, it is directly implied not only in those kinds of materialism which suppose that mental states and processes emerge from, or are constituted by, particular types or tokens of physical states and processes, but also in substance pluralist ontologies such as Cartesian substance dualism. In fact, supposing that different kinds of substances (e.g., physical vs. mental) exist independently of each other implies that they coexist in separate places.

Depending on the ontology that is being supposed, both the world wherein mental states and processes are supposed to have some particular place and the mental states and processes that are supposed to have some particular place in that world are conceived in a large variety of radically different ways. Nevertheless, these radically different conceptions have an important common feature. In fact, in order to localize mental states and processes in some particular place in the world, they must be conceived with reference to particular kinds of inner-world entities displaying particular kinds of states and processes. Accordingly, the mental states and processes that are supposed to have some particular place in the world must be conceived in terms of particular kinds objectively manifest states and processes, but they need not necessarily be conceived in terms of subjectively lived states and processes. This applies not only to mental states and processes in general, but also to consciousness in particular.

Depending on which particular kinds of inner-world entities are referred to in order to localize mental states and processes in some particular place in the world, the view that mental states and processes take place somewhere but not everywhere in that world may be, and has in fact been, specified in a large variety of ways. From a purely logical point of view, supposing nothing but an indefinite plurality of inner-world entities, the inner-world entities where mental states and processes might be supposed to take place may vary on two dimensions, and we will see that the inner-world entities where mental states and processes have in fact been supposed to take place do indeed vary on both dimensions:

On the one hand, mental states and processes might be supposed to take place in different kinds of *spatiotemporally exclusive* inner-world entities, i.e. in different kinds of *separate individuals*. For example, referring to well-established commonsense taxonomies of physical systems, they might be supposed to take place in different kinds of natural or artificial living or non-living systems such as human beings, animals, plants, or robots. In fact, while some philosophers and researchers seem to have ascribed mental states and processes, in general, and consciousness, in particular, only to human beings,[15] many others have ascribed it also to some other kinds of animals (for example to those capable of sophisticated, learnt, non-stereotyped

[15] The most famous examples of philosophers associated with this view are of course Descartes and "Cartesian" philosophers such as Malbranche, though at least in Descartes himself the issue is less clear than it is often taken to be (cf. Harrison 1992).

behaviors, or to those with a more or less complex nervous system),[16] and still others have ascribed it not only to some kinds of animals, but to virtually all kinds of natural living systems.[17] Furthermore, while some philosophers and researchers would ascribe mental states and processes, in general, and consciousness, in particular, only to some or all kinds of natural living systems but not to artificial or non-living systems, others would be ready to ascribe them also to these latter kinds of systems, and in particular to some kinds of artificial non-living systems, for example to artefacts which pass the "Turing test",[18] i.e. which pass "the empirical tests by which the presence or absence of [mental states and processes] is determined", or which fulfill the minimal requirements of the Integrated Information Theory (ITT) of consciousness.[19]

On the other hand, mental states and processes might be supposed to take place in different kinds of *spatiotemporally inclusive* inner-world entities, i.e. at *different system levels*. In particular, they might be supposed to take place not only at some given system level, i.e. in a given system as such, but also at lower system levels, i.e. in particular subsystems of that system, or at higher system levels, i.e. in superordinate systems of which that system is a subsystem. As a matter of fact, mental states and processes, in general, and consciousness, in particular, have frequently been localized not only in living systems such as human beings and other animals, but also and more precisely in particular subsystems of such systems (e.g., in the heart as the center of the circulatory system or in the brain as the center of the nervous system),[20] in particular subsystems of such subsystems (e.g., in the "pineal gland" or in particular structural or functional "brain networks"),[21] or even in the basic

[16] Cf. Moreno et al. (1997), Edelman and Seth (2009), and Griffin (2013) for recent examples.

[17] In recent times, this view has been suggested in particular by the Santiago School of cognition (Maturana and Varela 1980), on the one hand, and by the Adelaide school of cognitive biology (Lyon 2006; Lyon and Opie 2007) as well as by the Bratislava Center for Cognitive Biology (Kováč 2000), on the other. According to the Santiago theory of cognition, "[l]iving systems are cognitive systems, and living as a process is a process of cognition" and "this statement is valid for all organisms, with or without a nervous system" (Maturana and Varela 1980, 13). Similarly, according to the cognitive biology approach, every organism – whether mono- or multicellular – that can sense stimuli in its environment and respond accordingly is cognitive.

[18] Cf. Turing (1950). While the "Turing test", as originally conceived, was concerned with the question whether "machines [can] think" (Turing 1950, 433), an analogous test had been earlier conceived by Ayer (1936) in order to determine "the presence or absence of consciousness": "The only ground I can have for asserting that an object which appears to be conscious is not really a conscious being, but only a dummy or a machine, is that it fails to satisfy one of the empirical tests by which the presence or absence of consciousness is determined" (Ayer 1936, 140).

[19] The eminent neuroscientists Tononi and Koch (2015), for example, would go so far as to ascribe some minimal consciousness even to a "photodiode": "A corollary of ITT that violates common intuitions is that even circuits as simple as a 'photodiode' made up of a sensor and a memory element can have a modicum of experience", though "[i]t is nearly impossible to imagine what it would 'feel like' to be such a circuit" (p. 11). See Oizumi et al. (2014) for a detailed defense of this thesis.

[20] The view that not the brain but the heart is the "seat" of the mind was widespread in premodern thought (Cf. Lind 2007, and Clarke and O'Malley 1996) and has been defended by many premodern philosophers such as Aristotle (cf. Gross 1995).

[21] Notably, Descartes localized the mind in the "pineal gland" (cf. Descartes 1649, *The Passions of the Soul*, Art. 32ff), whereas contemporary philosophers and scientists generally localize mental

biological units of such subsystems of subsystems (e.g., in single neurons),[22] let alone in the basic physical constituents of such biological units and of matter in general (e.g., in atoms).[23] Somewhat less frequently, mental states and processes, in general, and (though less frequently) consciousness, in particular, have also been ascribed to superordinate systems that comprise "enminded" systems such as human beings and other animals as subsystems. For example, they have been ascribed to: (1) human-artefact systems, in so far as they are supposed to constitute "extended minds";[24] (2) organized groups of animals such as ant colonies, bee hives, insect swarms, bird flocks, fish schools, human communities, etc., in so far as they are supposed to constitute "group minds", "collective minds", or even "conscious communities";[25] (3) entire species of animals, in so far as they are supposed to be individuals with a mental life on their own;[26] and even (4) the Earth, other celestial bodies, and the universe as a whole, in so far as they are supposed to have a mental life on their own.[27] Thus, mental states and processes, in general, and consciousness, in particular, have been localized along the entire range of system levels, from the basic constituents of the world to the world as a whole, so that the view that mental states and processes take places only somewhere but not everywhere varies along virtually the entire arc of positions between particular kinds of "atomistic" pluralistic and "synechological" monistic versions panpsychism.[28]

Taken together, variations on these two dimensions, i.e. concerning different kinds of separate individuals, on the one hand, and concerning different kinds of system levels, on the other, result in a large variety of different ways of localizing

---

states and processes, in general, and consciousness, in particular, in functional "brain networks" (e.g., Laureys et al. 2016).

[22] Cf., for example, Hartmann (1869) and more recently Edwards (2005) and Sevush (2006, 2016).

[23] Cf., for example, Lotze (1852, 1856) and more recently Griffin (1998). This view has also been sympathetically discussed by Chalmers (1996) in the context of his attempt to tackle the "hard problem of consciousness".

[24] The "extended mind" hypothesis has been introduced by Clark and Chalmers (1998). Cf. Menary (2010) for an overview on the debate on this hypothesis.

[25] Cf., for example, Royce (1901, 1913) and more recently Wilson (2005), Couzin (2007, 2009), Tollefsen (2006), Theiner (2014), and Szanto (2014).

[26] Cf., for example, Fechner (1851, 1861) and Royce (1901, 1913). The view that species are individuals is not as far-fetched as it might seem at first sight, but is a respected hypothesis in biology that has been introduced by Ghiselin (1974) and Hull (1978).

[27] Cf., for example, Fechner (1851, 1861).

[28] The distinction between "atomistic", or "monadological", and "synechological" (from Greek *συνεχής* continuous, conjoined) versions of panpsychism has been introduced by Fechner (1855) and taken up by Hartshorne (1950): "The synechological view differs from the monadological view in that it does not associate the psychic unity with the single atoms, and thus does not suppose as many (conscious or unconscious) souls as there are metaphysically or physically discrete simple body-atoms in the world; it rather associates the psychic unity ultimately with the lawful connection of the whole system of the atoms of the world (God), thereby associating subordinate psychic unities (souls of human being and animals) with subsystems of this whole system" (Fechner 1855, 248f).

mental states and processes in particular places in the world. In fact, there seem to be only three positions on which virtually all these different views on the place of mental states and processes agree: first, (most) human beings have a mind on their own;[29] second, mere aggregates don't have a mind on their own, even if their components (e.g., atoms) do; and third, the distinction between entities that do and those that do not have a mind on their own is related to differences concerning their behavior and internal organization. Accordingly, the debate between different views eventually revolves around the question as to which kinds of behavior and internal organization legitimate to ascribe mental states and processes to a given entity, and the answers to this question vary in the degree to which these views abstract and generalize from the specificities of typically human behavior and internal organization. If this degree is low, an "anthropogenic" (rather than a "biogenic")[30] approach is adopted, and mind, in general, or consciousness, in particular, are ascribed for example only to those entities which show behavioral evidence of manipulating internal representations (e.g., to higher animals and to some kinds of artificial systems) or which have a more or less centralized nervous system (e.g., to higher animals, but not to artificial systems). By contrast, if this degree is high, a "biogenic" (rather than a "anthropogenic") approach is adopted, and mind, in general, or consciousness, in particular, are ascribed for example to all those entities which show behavioral evidence of sensory-motor coupling (e.g., to most animals and to some kinds of artificial systems) or which have an autopoietically evolved internal organization (e.g., to all living beings and at least potentially also to some kinds of artificial systems).[31]

3 A Tentative Account

The preceding brief and necessarily rough systematic overview of the basic kinds of answers that have been given to the question as to *where* mind takes place should have illustrated that this question has received an amazingly large variety of answers,

[29] Exceptions may be, for example, fetuses (e.g., Lagercrantz 2014) and individuals in certain clinical conditions such as persistent coma (e.g., Gosseries et al. 2014).

[30] The distinction between "anthropogenic" and "biogenic" approaches to cognition and mind has been introduced by Lyon (2006).

[31] Notably, the relationship between mind conceived in "anthropogenic" terms and mind conceived in "biogenic" terms is somewhat analogue to the relationship between respiration conceived as ventilation (which requires lungs and is behaviorally manifest as inhalation and exhalation) and respiration in general, including both physiological respiration (which may be realized by very different kinds of respiratory systems and may be behaviorally manifest in very different ways) and cellular respiration (which may be realized by respiratory mechanisms involving very different kinds of chemical substances). Thus, as already argued by Fechner (1861), showing behavioral evidence of mental representations and/or having a more or less centralized nervous system might be no more necessary for mind than inhalation and exhalation and/or having lungs are necessary for respiration.

ranging from virtually everywhere to virtually nowhere and from the one limit defined by places that don't contain other places (i.e., the places of "atoms", in the etymological sense of this term) to the other limit defined by the place that contains all other places (i.e., the all-encompassing place of the world). This amazingly large variety of answers hints at an intrinsic relationship between mind and place, which calls for an explanation. In particular, it needs to be explained why mind may be conceived, on the one hand, as the all-encompassing place in which everything takes place, and, on the other hand, as something which takes place in particular places, i.e. in some, every, or no inner-world entity. For obvious reasons, such an explanation cannot be given in terms of some particular ontological position, but it must be given in terms of a phenomenological approach, i.e. in terms of an analysis of the essential structures of phenomenal experience in which ultimately any meaningful and intelligible ontological claim must be grounded. In order to attempt such an explanation, it is therefore necessary to approach the relationship between mind and place from a phenomenological perspective, i.e. to ask where mind is supposed to take place when it is conceived from a phenomenological perspective.

From a phenomenological perspective, the all-encompassing place where everything takes place is the lifeworld. As such a place, the lifeworld is literally the *universe*, i.e. the unity that involves and contains any plurality, and the *cosmos*, i.e. the orderly system of whatever there is. In fact, from a phenomenological perspective, whatever there is, is somehow "given" in the lifeworld, i.e. it is part of the lifeworld as something which is, as it were, "perceived", "remembered", "imagined", "supposed" or otherwise "represented". From a phenomenological perspective, these "psychological" terms do not refer to different kinds of mental states or processes, but to different ways of "givenness", i.e. to different ways in which something may be part of the lifeworld. They acquire their psychological meaning only if the lifeworld is represented as "one's own" world and thus as a "subjective" world, i.e. as the world of a particular conscious subject among others coexisting in a supposed objective world. In fact, if the lifeworld is represented in this way, then it is supposed to be a place which is constituted by one's own mind in so far as it is supposed to vanish or change if one lost consciousness or if one's other mental states and processes vanished or changed (e.g., if one became blind, amnestic, or psychotic). Analogously, the same would be supposed to be the case for all other subjective worlds, i.e. for the worlds of all other conscious subjects which are supposed to coexist with oneself in the objective world.

From a phenomenological perspective, the concept of mind thus arises if the lifeworld is represented as a subjective world (i.e., the world of a particular inner-world entity) in opposition to other subjective worlds (i.e., the worlds of other inner-world entities) and to the objective world (i.e., the world in which these inner-world entities are supposed to coexist as conscious subjects). Accordingly, the phenomenological concept of mind implies an intrinsically ambiguous relationship between mind and world. In fact, mind, in general, and consciousness, in particular, is supposed to be: (1) that what constitutes a subjective world, and (2) something which takes place in particular kinds of inner-world entities, namely (a) in particular kinds of entities in any subjective world (i.e., in the central body that may be identified as one's own body, and in other, more or less similar bodies in its environment that may be identified as the bodies of other subjects), and (b) in corresponding particular kinds of entities in the objective world (e.g., in human beings, animals, etc.).

Now, as a subjective world is the place wherein takes place whatever there is for a particular subject, whereas the objective world is the place wherein takes place whatever there is in itself, independently of any particular subject, the intrinsically ambiguous relationship between mind and world implies a corresponding intrinsically ambiguous relationship between mind and place. In fact, mind, in general, and consciousness, in particular, may be supposed to be (1) that what constitutes the all-encompassing place of a subjective world, and (2) something which takes place in particular places in a) any subjective world, and b) the objective world. Furthermore, in virtue of this intrinsically ambiguous relationship, mind, in general, and consciousness, in particular, is often said, somewhat metaphorically, to be itself a place. On the one hand, in fact, it is often said to be the all-encompassing place in which takes place whatever there is for a particular subject; on the other hand, this all-encompassing place is at the same time often said to be an "inner" place with respect to the "outer" place of the objective world in which this subject is supposed to exist. For example, it is common to refer to that what there is for a particular subject as to something that is "in" that subject's consciousness, and thereby to consider it as something that cannot be observed by an "outside observer", but only by means of "introspection".

Starting from a phenomenological perspective, we thus encounter with respect to the concept of a subjective world an intrinsic ambiguity concerning the relationship between mind and world, and thus between mind and place, which is analogous to the ambiguity which we have encountered, from an ontological perspective, with respect to the variety of views on the objective world. In fact, as from an ontological perspective mind has been supposed to be on the one hand an all-encompassing place, in so far as it has been supposed to constitute the objective world, and on the other hand something which takes place in particular places, in so far as it has been supposed to take place in particular kinds of inner-world entities in the objective world, so from a phenomenological perspective mind, in general, and consciousness, in particular, may be, and in fact often is, said to be on the one hand a place wherein everything takes place, in so far as it is supposed to be that what constitutes a subjective world, and on the other hand something which takes place in particular places, in so far as it is supposed to be something which takes place in particular kinds of inner-world entities. Furthermore, as from an ontological perspective mind has been supposed to take place either everywhere, nowhere, or somewhere (but not everywhere) in the objective world, so from a phenomenological perspective mind, in general, and consciousness, in particular, may be said to take place either everywhere, nowhere, or somewhere (but not everywhere) in a subjective world: everywhere, in so far as it is supposed to be that what constitutes a subjective world and in so far as that what constitutes is involved in any part of that what it constitutes; nowhere, in so far as it is supposed to be that what constitutes a subjective world and in so far as that what constitutes is different from that what it constitutes; and somewhere (but not everywhere), in so far as it is supposed to be something which takes place in particular kinds of inner-world entities.

Given these analogies, it seems likely that the intrinsic ambiguity concerning the relationship between mind and world and thus between mind and place, which we have encountered, starting from a phenomenological perspective, with respect to the subjective world, is somehow at the basis of the analogous ambiguity which we have encountered, from an ontological perspective, with respect to the variety of views on the objective world. It remains however to be explored how a phenomenological approach characterized by that intrinsic ambiguity might explain not only the two general kinds of ontological positions represented by the two terms of the analogous ontological ambiguity: mind as place vs. mind in place,[32] but also the large variety of more specific ontological positions within each of these two general kinds. In the remaining part of this chapter, I will therefore roughly sketch what might be some basic results of such an exploration.

3.1 *Mind As Place*

There seem to be basically two different ways leading from the phenomenological view of mind, according to which mind is on the one hand that what constitutes any subjective world and on the other hand something that takes place in particular inner-world entities both in the subjective and in the objective world, to the ontological thesis that mind is the all-encompassing place wherein everything takes place. Given their point of arrival, both ways involve some kind of "dialectical" reasoning in the Kantian sense, i.e. they transcend the limits of possible experience.

The first way eventually amounts to conceiving the objective world in analogy to a subjective world: As any subjective world is supposed to be constituted by mind so that mind may be said to be the place wherein takes place whatever there is for a particular subject, so the objective world may be supposed to be constituted by mind so that mind is supposed to be the place wherein takes place whatever there is in itself, independently of any particular subject. This analogization, which extrapolates from what is supposed to constitute one's own and any other subjective world to what is supposed to constitute the one and only objective world, results in an idealist monistic ontology (cf. 1.2.) and it involves especially two variables which allow for different kinds of idealist monistic ontologies: the relationship between the world-constituting mind and the world that it is supposed to constitute, and the kinds of mental features that are supposed to constitute the objective world. Depending on whether the world-constituting mind is supposed to be different from or identical with that what it constitutes, there may be expected to be either theistic or non-theistic kinds of idealist monistic ontologies, and depending on which particular kinds of mental features are supposed to constitute the objective world, there may be expected to be different versions of either theistic or non-theistic kinds of

[32] For good reason, the title of this chapter, "The place of mind", is equivocal: it may refer either to mind as place or to mind as something that takes place in some place.

idealist monistic ontologies. Given that the move is from what is supposed to constitute any single subjective world to what is supposed to constitute the one and only objective world, the mental features that are supposed to constitute the objective world are likely to be those features that represent the literally least "subjective", i.e. the least passive and least receptive aspects of (human) subjectivity, such as thinking and willing, whereas more "subjective" aspects such as sensation and perception are likely to be either abstracted away or re-interpreted in active and non-receptive terms, e.g. as God's "intellectual intuition". This would explain why idealist monistic ontologies generally refer to the world-constituting "mind" in terms of an all-encompassing "spirit" or "will".

The second way eventually amounts to conceiving the objective world in analogy to those inner-world entities which are supposed to be enminded: As some inner-world entities are supposed to be enminded, so the objective world as a whole may be supposed to be enminded, and as those inner-world entities may be supposed not only to contain a mind, but to be pervaded by mind, so the objective world as a whole may be supposed not only to contain minds but to be pervaded by mind. This analogization, which extrapolates from what is supposed to be the case in one's own and any other subject's body to what is supposed to be the case in the objective world as a whole, results in a property-dualist monistic ontology (cf. 1.2.) and it involves especially two variables which allow for different kinds of property-dualist monistic ontologies: the kinds of mental features that are supposed to pervade the objective world, and the degree to which they are supposed to pervade it. Depending on the kinds of mental features that are supposed to pervade the objective world, some kinds of property-dualist monistic ontologies may be expected to refer to an all-pervasive reason, whereas others may be expected to refer to an all-pervasive feeling or will, and depending on the degree to which such mental features are supposed to pervade the objective world, some kinds of property-dualist monistic ontologies may be expected to be virtually indistinguishable from materialist ontologies, whereas others may be expected to be evidently different.

Given that these two ways have closely related points of departure and given that they both proceed by means of analogization, their points of arrival may be expected to be closely related as well. More precisely, the relationship between their points of arrival, i.e. between idealist monistic ontologies, on the one hand, and property-dualist monistic ontologies, on the other, may be expected to be analogous to the relationship between their points of departure, i.e. between what is supposed to constitute one's own and any other subjective world, on the one hand, and what is supposed to be the case in one's own and any other subject's body, on the other. In fact, as the difference between what is supposed to constitute one's own and any other subjective world, on the one hand, and what is supposed to be the case in one's own and any other subject's body, on the other, eventually amounts to the difference between two different perspectives on inner-world minds, so the difference between idealist monistic ontologies, on the one hand, and property-dualist monistic ontologies, on the other, eventually amounts to the difference between different perspectives on the supposed all-encompassing mind.

3.2 *Mind in Place*

The phenomenological view of mind, according to which mind is, on the one hand, that what constitutes any subjective world, and, on the other hand, something that takes place in particular inner-world entities both in the subjective and in the objective world, already implies the ontological thesis that mind takes place in particular places, for it supposes that mind takes place in particular entities in the objective world. More precisely, it implies that mind, in general, and consciousness, in particular, takes place in all those entities in the objective world which correspond to particular entities in a subjective world, namely to the central body that may be identified as one's own body, and to other, more or less similar bodies in its environment that may be identified as the bodies of other subjects. The phenomenological view of mind thus allows to explain the basic features of the large variety of answers to the question as to where, i.e. in which inner-world entities, mind takes place:

First, the phenomenological view of mind allows to explain why mind, in general, and consciousness, in particular, is generally supposed to take place in at least some kind of inner-world entities (e.g., in human beings). In fact, it implies that mind, in general, and consciousness, in particular, may be supposed to take place in one's own body and in other bodies of the same kind, thereby leaving open how this kind is to be specified. It therefore implies that mind, in general, and consciousness, in particular, may be supposed to take place in human beings, in so far as one identifies oneself primarily as a human being, but it does not exclude that mind, in general, and consciousness, in particular, may be supposed to take place in some other, either more or less specific kind of inner-world entities, if one identifies oneself primarily as an entity of that other kind. Accordingly, the phenomenological view of mind allows to explain not only why mind, in general, and consciousness, in particular, is generally supposed to take place in human beings, but also why mind, in general, and specific kinds of mental abilities (e.g., reason or consciousness), in particular, have sometimes (i.e., in some cultures and historical epochs) been supposed to take place either only in more specific kinds of inner-world entities (e.g., in human beings of a certain gender or race) or in less specific kinds of inner-world entities (e.g., in vertebrate animals).

Second, the phenomenological view of mind allows to explain why mind, in general, and consciousness, in particular, is more likely supposed to take place in some kinds of inner-world entities (e.g., in human beings and other "higher" animals) than in others (e.g., in "lower" animals, plants, or artefacts). In fact, it implies that mind, in general, and consciousness, in particular, is more likely supposed to take place in those bodies which are more "similar" to one's own body in phenomenologically relevant respects such as their form, development, and behavior.[33] In so far as one identifies oneself as a human being, it thus implies that mind, in general,

---

[33] Of course, the "similarity" in these respects also involves a "similarity" concerning the spatial and temporal scale of these respects. Cf. Morewedge et al. (2007) for empirical evidence of a "timescale bias in the attribution of mind".

and consciousness, in particular, is more likely supposed to take place in those inner-world entities whose form, development, and behavior are more "similar" to those of human beings. It therefore allows not only to explain why mind, in general, and consciousness, in particular, is generally more likely supposed to take place in human beings and other "higher" animals such as primates, mammals, or vertebrates, than in "lower" animals such as insects or sponges, or in plants or artefacts, but it also allows to explain why mind, in general, and consciousness, in particular, has traditionally been supposed to take place rather in human or animal bodies considered as a whole than in one of their body organs (e.g., in their brain).[34] Epistemologically, in fact, mind, in general, and consciousness, in particular, must have been supposed to take place in particular kinds of bodies, i.e. in particular kinds of directly observable systems, before it can be supposed to take place in particular kinds of body organs, i.e. in particular kinds of only indirectly observable subsystems. Only on this basis it is then possible to conclude vice versa from the presence of mind in particular kinds of subsystems to the presence of mind in systems with such subsystems.

Third, the phenomenological view of mind allows to explain why mind, in general, and consciousness, in particular, has been supposed to take place in a large variety of different kinds of inner-world entities, ranging from the largest (e.g., celestial bodies) to the smallest (e.g., atoms), but not in any inner-world entity whatsoever (e.g., not in mere aggregates such as heaps of sand, etc.). On the one hand, in fact, it implies that mind, in general, and consciousness, in particular, may be attributed, in principle, to any inner-world entity, for any inner-world entity is "similar" to any other inner-world entity in so far as they are all inner-world entities. On the other hand, however, it also implies that there must be some distinction between inner-world entities with and without mind, for otherwise there wouldn't be any reason to introduce the concept of mind at all. Now, as we have already seen above, the basic phenomenological criterion for distinguishing between inner-world entities with and without mind is their "similarity" to one's own body in phenomenologically relevant respects such as their form, development, and behavior, and thus, in so far as one identifies oneself as a human being, their "similarity" to human beings in those respects. From a phenomenological perspective, it may therefore be expected that mind is least likely attributed to those inner-world entities which are least "similar" to human beings in those respects. Accordingly, the phenomenological view of mind allows to explain why mind has occasionally been attributed to inner-world entities as different as atoms and celestial bodies, but virtu-

[34] Even today we would probably find it rather hard to attribute mental states to a brain in a vat, even if we were assured that the neuroelectric in- and output patterns are the same as in an embodied brain: to make such an attribution minimally plausible, we would probably need some concrete representation of the body-related stimuli and behaviors corresponding to those neuroelectric in- and output patterns. Vice versa, we would probably find it rather easy to attribute mental states to aliens if their behaviors were sufficiently similar to ours, even if their internal organization were completely unknown to us or radically different from that of known organisms on earth: to make us doubt in such an attribution, we would probably need to discover that their internal organization or their origin is rather similar to that of artefacts.

ally never to mere aggregates such as heaps of sand, etc. For despite the evident differences between them, atoms and celestial bodies share with human beings an important phenomenologically relevant characteristic that mere aggregates such as heaps of sand lack: their relatively autonomous and intrinsic unity.[35] In fact, while all inner-world entities, including mere aggregates, have some kind of unity in virtue of which they are distinct from each other, they differ in the way and in the degree in which they have some kind of unity, and *ceteris paribus* mind may be expected to be more likely attributed to those inner-world entities whose unity is more similar to that of human beings.

Finally, precisely because of its phenomenological character, the phenomenological view of mind conceives mind not only in terms of objectively manifest, but also in terms of subjectively lived states and processes, so that the previous three points apply to both objectively manifest and subjectively lived mental states and processes, in so far as they are linked. It therefore implies that both objectively manifest and subjectively lived mental states and processes take place in at least some (though not necessarily the same)[36] kinds of inner-world entities, without thereby implying in which precisely, and it does not exclude that both objectively manifest and subjectively lived mental states and processes take place in every inner-world entity, in so far as it does not exclude that mind is attributed to whatever are supposed to be the basic constituents of all inner-world entities. Accordingly, the phenomenological view of mind is consistent with virtually any ontology, including panpsychism, but not with eliminative materialism, in so far as that ontology holds that subjectively lived mental states and processes do not really exist.

4 Conclusion

In the history of philosophy and science, in general, and in contemporary philosophy and science, in particular, the question as to where mind takes place has received an amazingly large variety of answers ranging from virtually everywhere to virtually nowhere and from the one limit defined by places that don't contain other places (i.e., the places of "atoms", in the etymological sense of this term) to the other limit defined by the place that contains all other places (i.e., the all-encompassing place of the world as a whole). Despite their extreme differences, the various answers to this question are nevertheless related in so far as the entire range of these answers

---

[35] More precisely, it is their supposed autonomous and intrinsic unity what makes atoms and celestial bodies more likely candidates for the attribution of mind than mere aggregates such as heaps of sand. For as the example of celestial bodies illustrates, inner-world entities may become less likely candidates for the attribution of mind, if it is discovered that their unity is less autonomous and intrinsic than it seemed.

[36] In fact, the phenomenological view of mind does not exclude that some (though not all) objectively manifest mental states and processes may take place without being subjectively lived, but it excludes that subjectively lived mental states and processes take place without being somehow objectively manifest.

(though of course not the single, historically contingent answers themselves) may be accounted for in terms of the phenomenological view of mind. In fact, this view implies an intrinsic ambiguity concerning the relationship between mind and world and thus between mind and place, which allows to explain why on the one hand mind has been supposed to be the place wherein everything takes place, and why on the other hand it has been supposed to be something which takes place in particular places, namely in some, every, or no inner-world entity. In particular, this view allows to explain why mind has been supposed to take place in a large variety of different kinds of inner-world entities ranging from the largest (e.g., celestial bodies) to the smallest (e.g., atoms) and invariably including at least some kinds of human beings.

As our phenomenological experience of the world is the ultimate empirical basis on which any ontological concept must be ultimately grounded, so the phenomenological view of mind is the ultimate benchmark with which any ontological view of mind must be consistent. Given the epistemologically foundational character of phenomenological experience, it may well be that the "hard problem of consciousness" will never be solved. But to dismiss phenomenological experience for the sake of an ontological theory is a desperate and self-undermining enterprise that necessarily leads to an inconsistent and indeed literally "placeless" and "deplacing", i.e. disembodied and alienating view of mind.

References

Adams, F., and K. Aizawa. 2010. *The Bounds of Cognition*. Chichester: Wiley Blackwell.

Aristotle. 1961. *Parts of Animals*. Trans. A.L. Peck. Cambridge: Harvard University Press.

Ayer, A.J. 1936. *Language, Truth, and Logic*. London: Penguin.

Barrett, L. 2011. *Beyond the Brain: How Body and Environment Shape Animal and Human Minds*. Princeton: Princeton University Press.

Berkeley, G. 1710. *A Treatise Concerning the Principles of Human Knowledge*. Dublin: Aaron Rhames.

Bondi, H., and T. Gold. 1948. The steady-state theory of the expanding universe. *Monthly Notices of the Royal Astronomical Society* 108: 252–270.

Brüntrup, G., and L. Jaskolla. 2016. *Panpsychism: Contemporary Perspectives*. New York: Oxford University Press.

Chalmers, D. 1996. *The Conscious Mind*. Oxford: University of Oxford Press.

Churchland, P.M. 1981. Eliminative materialism and the propositional attitudes. *Journal of Philosophy* 78: 67–90.

Churchland, P.S. 1986. *Neurophilosophy: Toward a Unified Science of the Mind/Brain*. Cambridge: MIT Press.

Churchland, P.M. 1988. *Matter and Consciousness*. Rev. ed. Cambridge: MIT Press.

Clark, D.S. 2004. *Panpsychism: Past and Recent Selected Readings*. Albany: State University of New York Press.

Clark, A. 2008. *Supersizing the Mind: Embodiment, Action, and Cognitive Extension*. Oxford: Oxford University Press.

Clark, A., and D.J. Chalmers. 1998. The extended mind. *Analysis* 58: 7–19.

Clarke, E., and C.D. O'Malley. 1996. *The Human Brain and Spinal Cord: A Historical Study Illustrated by Writings from Antiquity to the Twentieth Century* (Second edition, Revised and Enlarged with a New Preface by Edwin Clarke). San Francisco: Norman Publishing.
Couzin, I.D. 2007. Collective minds. *Nature* 445: 715.
———. 2009. Collective cognition in animal groups. *Trends in Cognitive Science* 13 (1): 36–43.
Descartes, R. 1649. *Les passions de l'âme*. English edition: *The Passions of the Soul*. Trans. S. Voss. Indianapolis: Hackett Publishing 1989.
Edelman, D.B., and A.K. Seth. 2009. Animal consciousness: A synthetic approach. *Trends in Neuroscience* 32: 476–484.
Edwards, J.C.W. 2005. Is consciousness only a property of individual cells? *Journal of Consciousness Studies* 12: 60–76.
Fechner, G.T. 1851. *Zend-Avista: oder über die Dinge des Jenseits vom Standpunkt der Naturbetrachtung*. Hamburg: L. Voss.
———. 1855. *Über die physikalische und philosophische Atomenlehre*. Leipzig: Hermann Mendelssohn.
———. 1861. *Über die Seelenfrage*. Hamburg/Leipzig: Leopold Voß.
———. 1946. *The Religion of a Scientist. Selections from Gustav Theodor Fechne*. Trans. and Ed., W. Lowrie). New York: Pantheon.
Gallagher, S. 2005. *How the Body Shapes the Mind*. Oxford: Oxford University Press.
Ghiselin, M.T. 1974. A radical solution to the species problem. *Systematic Zoology* 23: 536–544.
Gosseries, O., H. Di, S. Laureys, and M. Boly. 2014. Measuring consciousness in severely damaged brains. *Annual Review of Neuroscience* 37: 457–478.
Griffin, David R. 1998. *Unsnarling the World-Knot: Consciousness, Freedom and the Mind-Body Problem*. Berkeley: University of California Press.
Griffin, Donald R. 2013. *Animal Minds*. Chicago: University of Chicago Press.
Gross, C.G. 1995. Aristotle on the brain. *The Neuroscientist* 1 (4): 245–250.
Harrison, P. 1992. Descartes on animals. *The Philosophical Quarterly* 42 (167): 219–227.
Hartsthorne, C. 1950. Chapter 35: Panpsychism. In *A History of Philosophical Systems*, ed. V. Ferm, 442–453. New York: The Philosophical Library.
Hegel, G.W.F. 1817. *Enzyklopädie der philosophischen Wissenschaften*. Heidelberg: Oßwald. English edition: *Encyclopaedia of the Philosophical Sciences in Basic Outline. Part 1: Logic*. Trans. and Ed., K. Brinkmann, and D.O. Dahlstrom. Cambridge: Cambridge University Press, 2010.
Herbart, J.F. 1828/1829. *Allgemeine Metaphysik, nebst den Anfängen der philosophischen Naturlehre*. Königsberg: August Wilhelm Unzer.
Hoyle, F. 1948. A new model for the expanding universe. *Monthly Notices of the Royal Astronomical Society* 108: 372–382.
Hull, D.L. 1978. A matter of individuality. *Philosophy of Science* 45: 335–360.
James, W. 1909. *A Pluralistic Universe: Hibbert Lectures at Manchester College on the Present Situation in Philosophy*. New York: Longmans, Green and Co.
———. 1911. Novelty and causation: The perceptual view (Chapter 13). In *Some Problems of Philosophy*, ed. W. James. New York: Longmans, Green & Co.
Kováč, L. 2000. Fundamental principles of cognitive biology. *Evolution and Cognition* 6 (1): 51–69.
Lagercrantz, H. 2014. The emergence of consciousness: Science and ethics. *Seminars in Fetal & Neonatal Medicine* 19: 300–305.
Laureys, S., O. Gosseries, and G. Tononi. 2016. *The Neurology of Consciousness. Cognitive Neuroscience and Neuropathology*. 2nd ed. Amsterdam: Elsevier.
Leibniz, G.W. 1714. Monadologie. English edition: *Leibniz's Monadology. A New Translation and Guide*, ed. L. Strickland. Edinburgh: Edinburgh University Press, 2014.
Lemaître, G. 1927. Un Univers homogène de masse constante et de rayon croissant rendant compte de la vitesse radiale des nébuleuses extra-galactiques. *Annales de la Société Scientifique de Bruxelles* A47: 49–59. English translation: Lemaître, G. 1931. Expansion of the universe. A

homogeneous universe of constant mass and increasing radius accounting for the radial velocity of extra-galactic nebulae. *Monthly Notices of the Royal Astronomical Society* 91: 483–490.
Lind, R.E. 2007. *The Seat of Consciousness in Ancient Literature*. Jefferson: McFarland & Comp.
Lotze, R.H. 1852. *Medicinische Psychologie, oder Physiologie der Seele*. Leipzig: Weidmann.
———. 1856. *Mikrokosmus. Ideen zur Naturgeschichte und Geschichte der Menschheit. Versuch einer Anthropologie*, vol. 1. Leipzig: S. Hirzel.
Lyon, P.C. 2006. The biogenic approach to cognition. *Cognitive Processing* 7 (1): 11–29.
Lyon, P.C., and J.P. Opie. 2007. *Prolegomena for a Cognitive Biology*. A conference paper presented at the Proceedings of the 2007 Meeting of International Society for the History, Philosophy and Social Studies of Biology, University of Exeter.
Mach, E. 1886. Beiträge zur Analyse der Empfindungen. Jena: G. Fischer. English edition: *The Analysis of Sensations and the Relation of Physical to the Psychical*. Trans. C.M. Williams. Chicago/London: The Open Court Publishing Company, 1917.
Maturana, H.R., and F.J. Varela. 1980. *Autopoiesis and Cognition. The Realization of the Living*. Dordrecht: Reidel.
Menary, R., ed. 2010. *The Extended Mind*. Cambridge, MA: MIT Press/Bradford.
Mesquita, B., L. Feldman Barrett, and E.R. Smith, eds. 2011. *The Mind in Context*. New York: Guilford Press.
Moreno, A., J. Umerez, and J. Ibañez. 1997. Cognition and life. The autonomy of cognition. *Brain & Cognition* 34: 107–129.
Morewedge, C.K., J. Preston, and D.M. Wegner. 2007. Timescale Bias in the attribution of mind. *Journal of Personality and Social Psychology* 93 (1): 1–11.
Oizumi, M., L. Albantakis, and G. Tononi. 2014. From the phenomenology to the mechanisms of consciousness: Integrated information theory 3.0. *PLoS Computational Biology* 10: e1003588. https://doi.org/10.1371/journal.pcbi.1003588.
Onions, R.B. 1951. *The Origins of European Thought: About the Body, the Mind, the Soul, the World, Time, and Fate*. Cambridge: Cambridge University Press.
Ramsey, W. 2016. Eliminative Materialism. In *The Stanford Encyclopedia of Philosophy*, ed. E.N. Zalta, (Winter 2016 Edition), URL: http://plato.stanford.edu/archives/win2016/entries/materialism-eliminative/.
Robbins, P., and M. Aydede, eds. 2009. *The Cambridge Handbook of Situated Cognition*. New York: Cambridge University Press.
Robinson, H. 2014. Substance. In *The Stanford Encyclopedia of Philosophy*, ed. E.N. Zalta (Spring 2014 Edition). https://plato.stanford.edu/archives/spr2014/entries/substance/.
Rosenberg, G. 2005. *A Place for Consciousness: Probing the Deep Structure of the Natural World*. Oxford: Oxford University Press.
Royce, J. 1901. *The World and the Individual*. New York: Macmillan.
———. 1913. *The Problem of Christianity*. Vol. II. New York: Macmillan.
Rupert, R.D. 2009. *Cognitive Systems and the Extended Mind*. Oxford: Oxford University Press.
Schaffer, J. 2016. Monism. In *The Stanford Encyclopedia of Philosophy*, ed. E.N. Zalta, (Spring 2016 Edition). http://plato.stanford.edu/archives/spr2016/entries/monism/.
Schopenhauer, A. 1819. Die Welt als Wille und Vorstellung. Leipzig: Brockhaus. English edition: Schopenhauer, A. 2010. *The World as Will and Representation*. Trans. J. Norman, A. Welchman, and C. Janaway. Cambridge: Cambridge University Press.
Seager, W., ed. 2018. *The Routledge Handbook of Panpsychism*. London/New York: Routledge.
Seager, W., and S. Allen-Hermanson. 2015. Panpsychism. In *The Stanford Encyclopedia of Philosophy*, ed. E.N. Zalta (Fall 2015 Edition). http://plato.stanford.edu/archives/fall2015/entries/panpsychism/.
Sevush, S. 2006. Single-neuron theory of consciousness. *Journal of Theoretical Biology* 238: 704–725.
———. 2016. *The Single-Neuron Theory. Closing in on the Neural Correlate of Consciousness*. London: Palgrave Macmillan.
Shapiro, L. 2011. *Embodied Cognition*. New York: Routledge/Taylor & Francis Group.

———., ed. 2014. *The Routledge Handbook of Embodied Cognition*. New York: Routledge.
Skrbina, D. 2005. *Panpsychism in the West*. Cambridge: MIT Press.
———., ed. 2009. *Mind That Abides: Panpsychism in the New Millennium*. Amsterdam: John Benjamins.
Spinoza, B. 1677. Ethica. English edition: *Ethics*. In *The Collected Writings of Spinoza*, ed. E. Curley. Princeton: Princeton University Press, 1985.
Sprigge, T. 1983. *A Vindication of Absolute Idealism*. London: Routledge and Kegan Paul.
———. 1998. Panpsychism. In *Routledge Encyclopedia of Philosophy*, ed. E. Craig. London/New York: Routledge.
———. 2007. My philosophy and some defence of it. In *Consciousness, Reality and Value: Essays in Honour of T. L. S. Sprigge*, ed. P. Basile and L. McHenry. Heusenstamm: Ontos Verlag.
Strawson, G. 2006a. Realistic monism: Why physicalism entails Panpsychism. In *Consciousness and Its Place in Nature: Does Physicalism Entail Panpsychism?* ed. A. Freeman. Imprint Academic: Exeter.
———. 2006b. Panpsychism? Reply to commentators with a celebration of Descartes. In *Consciousness and Its Place in Nature: Does Physicalism Entail Panpsychism?* ed. A. Freeman. Imprint Academic: Exeter.
Szanto, T. 2014. How to share a mind: Reconsidering the group mind thesis. *Phenomenology and the Cognitive Sciences* 13 (1): 99–120.
Theiner, G. 2014. Varieties of group cognition. In *The Routledge Handbook of Embodied Cognition*, ed. L. Shapiro, 347–357. New York: Routledge/Taylor & Francis Group.
Tollefsen, D.P. 2006. From extended mind to collective mind. *Cognitive Systems Research* 7 (2–3): 140–150.
Tononi, G., and C. Koch. 2015. Consciousness: Here, there and everywhere? *Philosophical Transactions of the Royal Society of London Series B, Biological Sciences* 370: 20140167. https://doi.org/10.1098/rstb.2014.0167.
Turing, A. 1950. Computing machinery and intelligence. *Mind* 59 (236): 433–466.
von Hartmann, E. 1869. *Philosophie des Unbewußten*. Berlin: Carl Duncker's Verlag.
Wilson, R.A. 2005. Collective memory, group minds, and the extended mind thesis. *Cognitive Processing* 6: 227–236. https://doi.org/10.1007/s10339-005-0012-z.
Wundt, W. 1863. Vorlesungen über die Menschen- und Thierseele, Hamburg: L. Voss. English edition: Wundt, W. 1894. *Lectures on Human and Animal Psychology*. Trans. J.E. Creighton, and E.B. Titchener. London: S. Sonnenschein.

Place and Positionality – Anthropo(topo)logical Thinking with Helmuth Plessner

Annika Schlitte

Abstract This paper explores a possible anthropological dimension of place by providing an interpretation of Helmuth Plessner's philosophical approach which proposes to understand it as a twofold "implacement" of man – discussing both *the place of man in the natural world* and *man's specific relation to place that makes him take his place in the natural world*. The interpretation follows Plessner's idea of a natural set of stages, developed in his major work *Die Stufen des Organischen und der Mensch*, leading from inanimate objects to plants, animals, and humans. According to Plessner, each stage differs from the other by virtue of its respective spatial delineation toward, and its position in, the world. With the paradoxical concept of man's "eccentric positionality", Plessner accounts for the exceptional position of man without thereby neglecting, on the one hand, his being part of nature, and, without falling prey to naturalism, on the other. Thus, Plessner offers an interesting perspective on place that also points beyond mere anthropocentrism.

Keywords Helmuth Plessner · Philosophical anthropology · Philosophy of nature · Place · Space · Boundary · Positionality

1 Prologue: Man and His Place in the Cosmos

In his book *The Wisdom of the World*, the French philosopher Rémi Brague starts from the observation that for a long period in Western thought the question of human nature was answered by reference to a cosmology that assigned him a place within an ordered whole (cf. Brague 2003). According to Brague, world and human subject were tied together by a shared moral order, defining distinct spheres which differed from each other in value. The basis for this model consisted in the assumption that

A. Schlitte (✉)
Institute of Philosophy, Johannes-Gutenberg University Mainz, Mainz, Germany

Interdisciplinary Research Group on "Philosophy of Place", Catholic University of Eichstätt-Ingolstadt, Eichstätt, Germany
e-mail: annika.schlitte@uni-mainz.de

T. Hünefeldt, A. Schlitte (eds.), *Situatedness and Place*, Contributions To Phenomenology 95, https://doi.org/10.1007/978-3-319-92937-8_8

the sublunary spheres were inferior to the superlunary ones, the former category therefore imitating the latter. Man, as a spiritual and sensible being, aspires to a place in the superlunary world, which is reflected by his upright position that lets him look to the sky. It is remarkable in this view of the world that the place of man in the cosmos cannot be seen independently of his value and his role in it: "his very place is enough to assign him a specific value" (Brague 2003, p. 105).

In modern times, this close relationship between ethics, anthropology and cosmology has changed with the advent of new astronomical knowledge. Astronomy no longer makes a distinction between the laws of the sublunary and the superlunary spheres. The new universe is morally indifferent and loses its role as a model, and, with the triumph of the biological theory of evolution in the nineteenth century, the estimation of man's position changes so fundamentally that not much is left of the ancient notion of a connection between the cosmological and the anthropological order.

I have made reference to Rémi Brague's philosophical remarks at the outset because he speaks of the 'place' of man in the world and thus reminds us of a long assumed connection between spatial categories and the normative order of the cosmos. Since we no longer share this view, this way of speaking about the "place" of man could be regarded as merely metaphorical, as an inacceptable anthropomorphism that inordinately expands the everyday experience of settling and living in place to a cosmological scale.

However, it may be that there is something in our concrete living in places that suggests drawing a connection to our human being in the world in general. One can find evidence for such a view in the phenomenological and hermeneutical tradition, for example in the works of Heidegger, who underlines the significance of dwelling for the human condition in his famous essay *Building Dwelling Thinking*, saying that "Dwelling, however, *is the basic character* of Being in keeping with which mortals exist" (Heidegger 1971, 158).

Although Heidegger himself describes the meaning of existence as primarily temporal in *Being and Time*, he attaches some importance to spatiality and place ("topos" or "Ort") in his later works (cf. Malpas 2012), as well as the whole phenomenological tradition comprises an intensive discussion on space, especially as "lived" space in contrast to the abstract account of space offered by modern natural sciences. Beyond the phenomenology of spatial experience, there is a line of thought which attributes an importance to place exceeding every analysis of lived space. In fact, the notion of place functions as a key term for some recent philosophical approaches, which have been labelled as a "philosophy of place" (cf. Malpas 1999; Casey 2009). The authors in question do not only stress the importance of "place" for our experience, but also its meaning as a philosophical core concept that has been neglected in modern philosophical thought. This is a perspective that allows for a possible connection between our concrete being in places and the place of man as such which is not merely metaphorical. In this view, the conditions of being human, our finitude, are not only expressed temporally as mortality, but also spatially as our being-in-place, as being in a space that opens up within boundaries.[1]

Undoubtedly, these approaches do not intend to develop an anthropology based on "place", as should be clear if we consider their strong reference to Heidegger.[2] The "philosophical topology" is therefore more a phenomenology or an ontology than, so to speak, an "anthropo-topology". Still, it is an interesting observation that philosophical anthropology in the twentieth century was aware of the importance of place, and reflected on man's relation to it. The importance of space and place is expressed in the idea that the exceptional position of man, if it is exceptional, has to be connected with his particular relation to space and place. How man relates to space and place is relevant for his position in the world as a whole.

2 The Situation of Philosophical Anthropology at the Beginning of the Twentieth Century

The following analysis of Plessner's basic concept of "eccentric positionality" is therefore based on the thesis that his answer to the anthropological question can be understood as a twofold "implacement" of man – as locating him, on the one hand, within a natural sequence of stages (*the place of man in the natural world*), and, on the other hand, to the extent that the specific spatiality of human existence itself is the basis for this situating act (*man's relation to place as something specific to him that makes him take his place in the natural world*). In this way, the question of living space conjoins the basic anthropological question of man and his "place in the cosmos" (Scheler 2009).

As the historical introduction above has shown, the answer to the anthropological question is not the same at every point in history but develops in contact with scientific knowledge about human beings. For modern philosophy, having left behind the certainties of the well-ordered Aristotelian cosmos or the Christian creation, the question of man's place in the world arises with new urgency. The safety in the center of the world offered by ancient thought is inaccessible to modern man, so much so that Georg Lukács at the beginning of the twentieth century could speak of a "transcendental homelessness" (Lukács 1971, p. 41). Man seems to be free to choose his place in the world, but he is also placeless insofar as his place in the world is not determined by his *nature* or *essence* – terms that have lost their persuasive power. The ancient concept of nature, referring to that which distinguishes one species from all the others, determined human good based on man's being

[1] For the connection between place and boundary cf. Casey's distinction between borders and boundaries in Casey 2017, pp. 7–27; Jeff Malpas also stresses the importance of boundaries for any philosophy of place (Malpas 2015, p. 122): "It is perhaps the attentiveness to the idea of boundary, and the necessity and productivity of that boundedness, that most clearly marks out the topographic or topological from other modes of contemporary thought."

[2] For Heidegger as a source for philosophical thinking on place cf. Malpas 2006.

determined by the nature of his kind, thus locating him in the cosmos. Against this, the modern concept of nature lacks any normative implications and subjects man to the same laws to which all natural beings are subjected. With the rise of the theory of evolution, which seems ultimately to deprive man of his exceptional position, this view became almost an inevitability at the beginning of the twentieth century.

The philosophical answer to the question of man is thus confronted with the dilemma of either assuming the naturalistic view of the empirical sciences, or of completely breaking away from the knowledge of the sciences. On this basis, there seems to be only the alternative of either reducing man himself to a natural thing or letting him fall completely out of the natural order.

This problem is addressed by philosophical anthropology, which can be understood as a quest for the place of man, which has become problematic after the end of faith in a comprehensive metaphysical order and the triumph of the evolutionary interpretation of man. At the same time, philosophical anthropology aims at a new approach in philosophy that avoids both the dangers of idealism and naturalism. Although it emerges under the premise of the scientific view of man, it cannot fully be understood as a search for the attributes of *homo sapiens*, but rather as a self-reflection on the identity of the living being who philosophizes about it.

While Max Scheler will take the path of defining man as a spiritual being, and in so doing delimits human beings clearly from the world of animals (cf. Scheler 2009), Plessner does not approach man from a theory of spirit, but in the direction of a philosophy of nature. At the same time, he starts with a concept of life that does not belong to the philosophy of life in the strict sense. In what follows below, his conception is examined in greater detail with regard to its reflections on the spatial dimension of human life.

3 Plessner's Anthropological Approach as an "Implacement" of Man

In his outline of a philosophical anthropology in *The Levels of the Organic and Man* from 1928 (Plessner 1975), Plessner attempts to include the natural and the spiritual aspect of man in his theory, taking them as united in experience and starting from the concept of life.[3] In contrast to Scheler, he constructs his anthropology on a philosophy of nature. In the first chapter, his program is announced as follows:

> Without philosophy of man, no theory of human experience of life in the humanities. Without a philosophy of nature, no philosophy of man. (Plessner 1975, p. 26)[4]

[3] For an overview of Plessner's work in English see the collection of essays in De Mul 2014a., esp. the introduction (De Mul 2014b), on the German discussion see Köchy and Michelini 2015, and, most recently, the special issue of the *Zeitschrift für Kulturphilosophie* (Schürmann 2016).

[4] All quotes from Plessner are my translation, if not otherwise indicated.

Although he aims at a definition (*Bestimmung*) of man, he does not open his analysis with the internal perspective, rejecting the idea of introspection as the privileged access to philosophical knowledge. In Plessner's theory, man is addressed neither from the perspective of the subject nor as the object of science, but as "object and subject of his life, i.e. in the way he is at the same time object and center for himself" (Plessner 1975, p. 31). In the preface to the first edition, he also presents his position as distinct from Heidegger's approach of an existential analysis taking "Dasein" as a starting point (Plessner 1975, p. V).[5] What Heidegger underestimates according to Plessner, is the fact that the human being-in-the-world cannot be understood independently of its relation to the way in which other natural beings are in the world, and thus he seems to begin at the wrong end when he starts with an analysis of "Dasein".[6] By contrast, Plessner outlines a sequence of stages from the inorganic to the organic, differing by their respective spatial delineation toward the world. Thus, the analysis does not start with a description of human beings but with the thing and its limits in space.

3.1 *From the Double Aspectivity of the Physical Thing to the Boundary of the Organic*

The structure of the argument Plessner develops in his book is always the same: He identifies a certain structure, a logic, which evolves at one stage and becomes thematic at the next, so that the characteristics of the lower stages are still present but reach a different level of consciousness and importance. Finally, he devotes a great part of his investigation to the distinction of living from merely natural beings. Following the assumption that the features characterizing the lower stages do not simply disappear in the higher stages, but are preserved and overcome ("*aufgehoben*"), Plessner's approach to man does not leave the sphere of a philosophy of nature.

The first point he mentions in his description of non-living beings, is the so-called "double aspect" ("*Doppelaspekt*", Plessner 1975, p. 80). In perception, the physical thing already features a certain "double aspectivity", which consists in its appearing as a core with properties belonging to it (cf. Plessner 1975, p. 81).

As Plessner describes, we perceive the green color of the leaf as belonging to the leaf and do not say, conversely, the green owned the leaf. This circumstance indicates the "dependence of the property on the core substance of the thing" (Plessner 1975, p. 82). The substantial core ("*Kern*") of a perceived object, however, stays concealed inside, in contrast to its external, phenomenal properties, and thus the perceived object always has a certain dimension of depth.

[5] On the relation between Heidegger and Plessner cf. the remarks in De Mul 2014a, p. 15, also Köchy and Michelini 2015.

[6] Nevertheless, there are not only clear influences of Heidegger on Plessner (also in terms of language), Heidegger has also turned his attention to the question of life and man's relation to animals and inorganic entities (cf. Heidegger 2001).

The dyad of a core and properties is supplemented by a fundamental "aspectivity" of every perceived thing (Plessner 1975 p. 83), which lets us perceive only one side of a thing at a given moment – thus the dimension of depth is complemented by the plurality of sides. Here, the relation to Husserl's "adumbrations" ("Abschattungen") is obvious.

What we perceive is always only one aspect of the perceived thing, but this does not imply a subjective perspective in the sense that everyone has his or her own perspective on the object. It means rather a potentiality of the object, or, as Plessner says, "the possibility of being opposite to the subject, guaranteed by the appearance" (Plessner 1975, p. 83).

What Plessner describes in spatial terms as a relation between different sides of the thing is not really a spatial relation, as he shows by a comparison with elements of the spiritual life, which show similar structural aspects of depth and perspective without being spatial. For example, I may have a feeling without recognizing all dimensions and implications of it at the same time. If we speak of two sides of a problem, we usually do not assume that these "sides" are spatial in the same sense as two sides of a coin. Even with regard to physical objects, the philosophical analysis reveals that we have to be careful with the spatial terms involved in their description. As to their "double aspectivity", Plessner explains that the "core" is not the same as a spatial center:

> One can put one's finger on the center and the sides in a spatial sense. On the center and sides as thing-constitutive characters, however, one cannot. (Plessner 1975, p. 84)

One cannot find the "core" of a perceived object by cutting it open and penetrating into its center. For this reason, we are dealing with "space-like" ("*raumhaft*") but not "spatial" ("*räumlich*") characters (Plessner 1975, p. 85), as Plessner explains, although we first encounter this difference in the perception of spatial phenomena. Plessner now uses this concept of double aspectivity in order to preserve the continuity between the inorganic and the organic entities on the one hand, and, on the other hand, to explain the overcoming of the inorganic stage.

For the organic being also owns this double aspect, realized as the internal and the external, but in it, this double aspect itself becomes apparent as an essential property of the living being. Plessner states that the "physical things of the intuition, in which a principally divergent outside-inside-relation occurs objectively as belonging to their being, are called living" (Plessner 1975, p. 89), and he draws a connection between this aspect of being determined through the inside-outside-divergence with the occurrence of a *boundary* belonging to the living body. As described above, the divergence between core and properties is not equivalent with the spatial appearance of surface and interior in a physical body. In the case of a living body, by contrast, the spatial contour and the irreducible divergence between inside and outside coincide in its boundary. Living beings do not only possess a boundary, they *are* their boundary, so to speak. The living being therefore does not fall prey to Cartesian dualism: It is not inside or outside, but it shows itself precisely in the realization of the transition between both, as a realization of a boundary.

The physical thing is separated from other bodies, but its boundary is not part of itself. Non-living bodies have no boundary in the sense of a "skin", enabling a transition from the inside to the outside. The boundary between two physical bodies is situated just in the gap between them, but has nothing to do with them. They are located in a relative space, where one point is as good as the other, but they do not have a relation to their environment. Living beings, on the other hand, "have a phenomenal, intuitive boundary (*"erscheinende, anschauliche Grenze"*, Plessner 1975, p. 100). Through this boundary, however, it is possible for them to come into contact with their environment, and they thus acquire a completely different relationship to space.

3.2 From Boundary to Positionality

In general, all natural bodies are in one place in space and time. Spatial-temporal binding to a particular place makes a body identifiable and accessible to the empirical sciences. In contrast to inorganic objects, however, living bodies are characterized by the fact that they can have a relation to the place where they are located as physical bodies.

This is possible by the specificity of the boundary, which not only closes the living body inwards, but also opens it to the outside. Insofar as living beings can open themselves outwardly, they are also able to transcend their physical body and to relate to it. With this, however, a completely new position on the world is reached, which Plessner describes as "positionality" (Plessner 1975, p. 127).

The double aspect of inside and outside, which consists of the core and its properties in the physical thing, thus becomes itself an essential property of living beings. Living beings *have* an inside and an outside. Through its boundaries, which open it to the outside, a living being not only *is*, but it *is beyond itself* ("*über sich hinaus*"), through the boundary that closes it inwardly, it is also *into itself* ("in sich hinein"), as Plessner says (cf. Plessner 1975, p. 128 f.).

Thus a living being is not just something that simply occurs in the world, but it is "placed" in this position of the *beyond* and *into* itself. The essence of positionality therefore means:

> In its vitality, the organic body differs from the inorganic through its positional character or its positionality. Under this, I understand the basic trait of its essence that makes a body in its being a posited one. (Plessner 1975, p. 129)

This being placed or being posited – a conversion of the idealistic "positioning" ("*Setzung*") through the I – makes it possible for a living being to take a certain distance from itself in respect of its belonging to the stage of natural beings, while, nevertheless, the features of natural bodies still pertain to it.

> It belongs to its positional character that a thing is beyond itself, into itself. In order to account for this claim, the thing has to be enabled to distance itself from itself. (Plessner 1975, p. 132).

For Plessner, it is precisely the possibility of distancing oneself from oneself as a natural body that makes an open relation to space possible. Insofar as the living being can dissociate itself from itself as a body, it enters into an interrelation with the environment. As the concept of positionality already implies, Plessner describes the relation of living beings to the world as a special form of spatial relation. In a later text, Plessner makes this connection between positionality and spatial relations very clear:

> In its vitality, the organic body differs from the inorganic in that it has positionality, i.e. in that it is distinguished from its environment and has a different relationship to its boundary, depending on the organizational level. A living being is placed not only *in* its environment, but also posed *against* it. It lives in dynamic relation both *to* its environment and, on the contrary, *back* to itself, the living thing; that is, in the *double aspectivity* of untranslatable directional opposites. [...] This applies to all living things: plants, animals and humans. (Plessner 1982, p. 9)

In order to elaborate the specific spatial relation of the living, Plessner distinguishes – similar to the phenomenological tradition – between the measurable space of physics and a lived space, "*der lebendigen Evidenz unmittelbarer Anschauung*" (Plessner 1975, p. 181), in which no measurable quantities, but rather experienced qualities, form the structuring principles. With regard to this qualitative space, there are remarkable differences between organic and non-organic beings:

> If we consider [...] an inanimate physical body in its relation to space and time as forms of being that can only be conceived in terms of experience, that is, in its intuitive relation to besides, over, behind, now, then, back then, etc., an essential indifference of its nevertheless spatially and temporally 'determined' being toward its position and duration becomes apparent. (Plessner 1975, p. 182)

The relation to space that living bodies have looks quite different. In contrast to non-living things, a living being does not simply fill a site ("Stelle") in space, "but it has a place, or more strictly, it asserts from itself a place, its 'natural place'" (Plessner 1975, p. 131). The specific relation living beings have to space also manifests itself with regard to time, by becoming and passing away, involving a process of growth and aging. Inasmuch as spatiality and temporality belong to the self of a living being, it separates itself from the merely relative relation of natural bodies to one another, in the way that its own place and its existence in time becomes an absolute reference point:

> The living body is not in this sense indifferent toward space and time: it grows and it ages. Its being-anywhere, its becoming-changed-at-sometime, however, may have the merely relative meaning of determinedness, as is essential to all bodies. However, while these consist in their position measured according to the coordinates of space and time [...], living things are related to their place in space and time. The growing body has an absolute measure of space in the growth of its boundaries, an absolute measure of time in its aging. (Plessner 1975, p. 182)

Space and time are therefore nothing external to living beings, but "forms of space and time move from the position of conditioning external forms into the position of 'conditioned' inner characters of being" (Plessner 1975, p. 183). So living beings can have a kind of "natural place" – something Aristotle had posited for all things (cf. Plessner 1975, p. 184). Plessner now describes how far this positional character is prefigured in plants and realized in animals.

3.3 *From Positionality to Centric and Eccentric Positionality*

In order to differentiate between plants and animals, he distinguishes between the "open form" of the plant and the "closed form" of the animal. While the organism of a plant is directly integrated into its surroundings and constitutes only a dependent section of the surrounding circle of life ("*Lebenskreis*", cf. Plessner 1975, p. 219), the organism of the animal in its closed form is only indirectly integrated and thus counts as an independent section of its "*Lebenskreis*" (cf. Plessner 1975, p. 226). Different from a plant, the animal has consciousness, albeit not self-consciousness, but a consciousness that creates a distance to the physical body. This lets the animal make a separation between its own inside and its surroundings. In itself, it has a center that is missing in the plant.

In its own standpoint, the animal has an absolute "here" which cannot be relativized at all (cf. Plessner 1975, p. 238 f.). The animal stands against its environment and distances itself from its own physical body, but it has no consciousness of itself. Its special position is now called "centric positionality" by Plessner:

> The animal lives out of its midst, into its midst, but it does not live as a midst. It experiences content in an environment, strange and its own, it is also able to gain control over its own body, it forms a system which is reflexive, but it does not experience – itself. (Plessner 1975, p. 288)

To experience oneself as oneself is reserved for humans, who, as eccentric ("*ex-zentrisch*") positioned beings, maintain a different relation to the world than the other living beings, while the characteristics described above – "double aspectivity", "boundary", "positionality", "centering" still remain valid for humans.

> Man as the living thing that is placed in the midst of his existence, knows this midst, experiences it, and is thus beyond it. He experiences the bond in the absolute here-and-now, the total convergence of the environment and his own body against the center of his position, and is therefore no longer bound by it. [...] If the animal is centric, man's life is eccentric, without being able to break the centering at the same time. Eccentricity is the characteristically human form of its frontal being posited toward the environment. (Plessner 1975, p. 291f.)

It is extremely important for Plessner to describe man as the highest stage in an organic process of development without, at the same time, letting him lose his connection to organic life. Bodily, he stays an animal, and by virtue of his eccentricity he accomplishes something already inherent in the animal as the first step in the distancing from bodily being. Eccentricity is therefore no new organic form (cf. Plessner 1975, p. 293), but the becoming-reflexive of the former stage.

By being able to distance himself even from his inner center, man becomes "ex-centric". He steps out of himself and opposite to himself ("*aus sich heraus und sich gegenüber*"). Related to this ability is the fact that his relation to space and place also becomes a different one.

> As I, which enables the full turning of the living system to itself, man no longer stands in the here-and-now, but 'behind' it, behind himself, place-less, in nothingness, he merges with nothingness, in the space-time-like nowhere-no-time. Placeless-timeless, he makes possible the experience of himself and at the same time the experience of his placelessness and timelessness as standing outside of himself, because man is a living thing, which is no longer standing only in itself, but whose standing-in-itself becomes the foundation of his standing. He is placed in his boundary and therefore beyond it, which limits him, the living thing. He does not only live and experience, but he experiences his experience. (Plessner 1975, p. 292)

With eccentric positionality, Plessner describes a characteristic of man, which recalls the self-reflectivity idealistic models are based on, but he approaches this aspect not from the inner perspective of introspection, but from the external perspective of spatial relations in nature.

3.4 From Nature to Culture

From the eccentric position of man toward himself, his particular relation to the world arises, which Plessner describes as consisting of the three spheres "external world", "internal world", and "with-world" (cf. Plessner 1975, p. 293). The three "worlds" are not reducible to each other, and each of them entails a specific relation to space.

Man's relation to the external world is characterized by his being a living body ("Leib") on the one hand, and a physical body ("Körper") on the other. Both the characterizations of space belonging to the living body (up-down, front-back, right-left, etc.) and those belonging to the physical body (arbitrary points or "sites" in an isotropic spatial continuum) work together in man's relation to external space and converge only in the paradoxical structure of eccentricity that challenges both concepts of space – pointing towards the "I" of self-consciousness, which is neither identical with the physical nor with the lived body (cf. Plessner 1975, p. 294 f.).

The internal world is the world inside the lived body, where the living being has a feeling for itself as opposed to the outside world. According to his eccentricity, man is never fully involved in the inner world and does not completely identify with the "here-now", as the animal does. As eccentric, man is at the same time in his place and dis-placed.

Moreover, he is also part of a third world, the "with-world". This with-world transcends the absolute here of the individual and leads to a kind of common place, in which the different individual "here-nows" coincide in a shared sphere, "where every human being is standing where the other is standing" (Plessner 1975, p. 304). This sphere is referred to as "spirit" ("Geist") by Plessner.

From the eccentricity of man, however, the practical question arises how man can cope with his existence (Plessner 1975, p. 309), for he is by no means fixed on anything: "As eccentrically organized being, he has to make himself that which he already is." (Plessner 1975, p. 309) He cannot directly follow a given direction or *telos*, rather he must conduct his life (cf. Plessner 1975, p. 310). Plessner regards the question of the good life as "the typical expression of the broken and eccentric character, which no naive, natural, unbroken, ephemeral and tradition-bound epoch of mankind could escape" (Plessner 1975, p. 309). In this context, he also speaks of the "constitutive homelessness of the human being" (Plessner 1975, p. 309). Since the structure of human life implies this lack of orientation, cultural activity appears as a "logical" consequence of the natural conditions.

Plessner now derives three anthropological principles from this state of man. The first principle is the "law of natural artificiality" ("*natürliche Künstlichkeit*", Plessner 1975, p. 309). By lacking a natural equilibrium, man is naturally designed to create an artificial substitute – culture. Here we find motifs that pervade the entire anthropology since the myth of Prometheus – man as a deficient being ("*Mängelwesen*", Herder) or culture as compensation (Plato), but with the difference that the necessity of culture is now something that appears within the natural development, in a paradoxical structure that calls for its transcendence. It is interesting that Plessner also explicitly links this deficiency to man's lack of place at this point:

> The constitutive lack of equilibrium of his particular mode of positionality – and not the disturbance of an originally normal, once harmonious and now again being able to become harmonious system of life – is the 'cause of culture'. Being existentially indigent, naked and 'halved', man's artificiality is the expression of his nature. It is the detour to a second fatherland set by eccentricity, in which he finds his home and absolute rootedness. Placeless, timeless, set into nothingness, the eccentric form of life creates its own ground. Only if he creates it he has it, and is borne by it. (Plessner 1975, p. 316)

The second law is that of "mediated immediacy" ("*vermittelte Unmittelbarkeit*", Plessner 1975, p. 321), which states that man has only indirect access to the world. As being essentially expressive, man produces cultural objects, symbolic contexts of meaning, and history.

Finally, the third law is that of the "utopian standpoint" (Plessner 1975, p. 341), which again refers explicitly to spatiality. The concept, which is paradoxical in itself, points to the striving of man for an implacement despite his ex-centric lack of place – for wholeness and a metaphysical "home". At this point, Plessner comes to speak of religion, which can give man what nature and mind must deny him, since his natural character is a broken, searching one:

> Ultimate attachment and integration, the place of his life and death, reconciliation with fate, interpretation of reality, home ["Heimat", A.S.] is given only by religion. (Plessner 1975, p. 342)

The eccentric positional form thus seems to demand an absolute being as a correlate, but for Plessner the eccentricity of the human being, in its inconsistency, also points beyond the possibility of such a reconciliation.

4 Anthropological Consequences

This journey through Plessner's major work in anthropology has shown that his special relation to space is decisive for man's position in the stages of the organic. Whereas the inanimate body is totally indifferent toward space, plants and animals are in a constant relation to it. Man, however, can distance himself from it, so that his relation to space becomes a disrupted one. One can thus, as below, represent the series of stages also with regard to space:

Physical things	Plants	Animals	Humans
			As self-consciousness transcending the centricity at a place, place-less, ex-centric
		As having consciousness centered in space and place, here-now	As having consciousness centered in space and place, here-now
	As a living being in relation to space and place through the boundary	As a living being in relation to space and place through the boundary	As a living being in relation to space and place through the boundary
As a natural body having a double aspect, without an intrinsic relation to space and place	As a natural body having a double aspect, without an intrinsic relation to space and place	As a natural body having a double aspect, without an intrinsic relation to space and place	As a natural body having a double aspect, without an intrinsic relation to space and place

In the paradoxical formulation of an eccentric positionality, which nevertheless includes centering, the problem of the human relation to space is revealed. Neither is man completely bound to a certain place, nor is he able to take leave of any place where he or she is, and to be taken up entirely within an abstract sphere removed from space and time. Rather, the stage of animal centering dwells in him as a longing, which, however, clearly predisposes him to the equally paradoxical utopian position, which points beyond any concrete location and makes man search for a "metaphysical" home without finding one.

The intermediate, the "halfness" ("*Hälftenhaftigkeit*") of his existence, which characterizes his position within the order of the organic as a whole, thus has its correspondence at the level of the spatial. Here, too, man is in transition, in movement from the attachment to a "natural place" to the absolute "placelessness".

If we take this relationship between the anthropological place and the analysis of experienced space seriously, it is only logical to examine the spatial and temporal factors of our existence as our forms of being-in-the-world. Space can reveal the special position of man, which Plessner still assumes, even if he does not exclude man from the natural order. One advantage of Plessner's approach is that, by avoiding a strict dualism, it allows the different levels of human life to be included, since neither the physical space, the space of perception, nor the space of action alone make up the specific spatial reference of man, but a combination of them.

Plessner's approach also makes it possible to develop a theory of space that is not anthropocentric from the outset, drawing attention also to non-human forms of spatial reference and including them into the analysis of human space. Plessner, however, remains sensitive to the peculiarity of man, and does not fall into a hasty naturalism, which reduces man to his physicality. The fact that his spatiality can also provide access to the special position of man is, however, a notion that Plessner reaches precisely because he leaves the familiar terrain of introspection and proceeds from a philosophy of nature. Thus, space and place are not merely external magnitudes, but belong to our access to the world, and determine them essentially – without being located merely inside the subject. They become fundamental conditions of human life, while at the same time they are "lived" and "experienced" in concrete situations and actions. In this way, an anthropological dimension of space is opened, which is worthwhile to be further elaborated upon in the context of philosophical self-reflection.

References

Brague, R. 2003. *The Wisdom of the World*. Chicago: The University of Chicago Press.

Casey, E. 2009. *Getting Back into Place*. 2nd ed. Bloomington: Indiana University Press.

———. 2017. *The World on Edge*. Bloomington: Indiana University Press.

De Mul, J. 2014a. Artificial by Nature. An Introduction to Plessner's Philosophical Anthropology. In de Mul (pp. 11–37).

———., ed. 2014b. *Plessner's Philosophical Anthropology. Perspectives and Prospects*. Amsterdam: Amsterdam University Press.

Heidegger, M. 1971. Building dwelling thinking. In: *Poetry, Language, Thought*. Trans. Albert Hofstadter (pp. 141–159). New York: Harper & Row.

———. 2001. *The Fundamental Concepts of Metaphysics: World, Finitude, Solitude*. Trans. William McNeill and Nicholas Walker. Bloomington: Indiana University Press.

Köchy, K., and F. Michelini, eds. 2015. *Zwischen den Kulturen. Plessners "Stufen des Organischen" im zeithistorischen Kontext*. Freiburg: Alber.

Lukács, G. 1971. *The Theory of the Novel. A Historico-Philosophical Essay on the Forms of Great Epic Literature*. Cambridge, MA: MIT Press.

Malpas, J. 1999. *Place and Experience*. Cambridge: Cambridge University Press.

———. 2012. *Heidegger and the Thinking of Place: Explorations in the Topology of Being*. Cambridge, MA: MIT Press.

———. 2006. *Heidegger's Topology. Being, Place, World*. Cambridge, MA: MIT Press.

———. 2015. Self, other, thing: Triangulation and topography in post-Kantian philosophy. *Philosophy Today* 59: 103–126.

Plessner, H. 1975. *Die Stufen des Organischen und der Mensch*. Berlin/New York: De Gruyter.

———. 1982. Der Mensch als Lebewesen. In *Mit anderen Augen. Aspekte einer philosophischen Anthropologie*, 9–62. Stuttgart: Reclam.

Scheler, M. 2009. *The Human's Place in the Cosmos*. Trans. Manfred S. Frings. Evanston: Northwestern University Press.

Schürmann, V. (ed.). 2016. Special Issue: Helmuth Plessner. *Zeitschrift für Kulturphilosophie* No. 2 (2016).

Birth in Language: The Coming-to-Language as a Mark of Non-difference in Gadamer's Hermeneutics

Tsutomu Ben Yagi

Abstract This essay advances an interpretation of Hans-Georg Gadamer's hermeneutics that is worked out in light of the works of two contemporary German philosophers, Bernhard Waldenfels and Peter Sloterdijk. By bringing Gadamer in dialogue with these two thinkers, I aim to present the dimension of non-difference that characterises his hermeneutic thinking. To this end, Gadamer's views on the mother tongue (*Muttersprache*) and the home (*Heimat*) will be explored by examining the phenomena of foreignness and birth, each of which was developed by Waldenfels and Sloterdijk respectively. In the first part of the essay, I offer a reading of Gadamer that shows how Waldenfels' critique of Gadamer based on radical foreignness is unfounded and misguided. I then proceed to make the case that the mother tongue must be thought in terms of the coming-to-language (*Zur-Sprache-Kommen*), the notion of which is employed by both Gadamer and Sloterdijk. As a notion that marks our original belonging to the world, I contend that the coming-to-language needs to be understood outside of the logic of difference. This essay thus concludes with the reflection that hermeneutics is essentially bound up with the character of non-difference, which is to be distinguished from the thinking of difference that dominates contemporary discourse.

Keywords Hans-Georg Gadamer · Peter Sloterdijk · Bernhard Waldenfels · Hermeneutics · Language · Non-difference · Birth · Coming-to-language

---

An earlier draft version was presented in Roanoke, U.S.A. on 12 September 2014 at the 9th Annual Conference of the North American Society for Philosophical Hermeneutics (NASPH). A reworked version was then presented in Johannesburg, South Africa on 28 March 2015 at the 3rd International Annual Conference of the Centre for Phenomenology in South Africa (CPSA).

T. B. Yagi (✉)
Catholic University of Eichstätt-Ingolstadt, Eichstätt, Germany

T. Hünefeldt, A. Schlitte (eds.), *Situatedness and Place*, Contributions To Phenomenology 95, https://doi.org/10.1007/978-3-319-92937-8_9

1 Introduction

The aim of this paper is to take up and address two distinct questions that have arisen in recent philosophical discussions in Germany from the perspective of Gadamer's philosophical hermeneutics: namely, the questions of foreignness (*Fremdheit*) and birth (*Geburt*). As two seemingly unrelated phenomena, these questions have emerged independently of one another, insofar as the former question has been posed and developed most notably by Bernhard Waldenfels, whereas the latter has been brought forward by Peter Sloterdijk.[1] In approaching these questions from the perspective of Gadamer's hermeneutics, I wish to re-present and situate his thought within these recent discourses. I believe such a task is called for, for Gadamer's hermeneutics, though to a different degree, is received critically by the two thinkers just mentioned. Yet, at least with regards to the thought of Sloterdijk, there are many respects in which Gadamer developed and shared a similar line of thought, especially when one attends to Gadamer's later writings. I shall therefore attempt above all to respond to the challenge brought forth by Waldenfels by identifying a moment of non-difference in both Gadamer and Sloterdijk.[2] That is to say, I wish to develop the thesis that the difference between the own (*Eigene*) and the foreign (*Fremde*) follows from – and does not precede – the original moment of our birth, the moment of our coming-to-the-world (*Zur-Welt-Kommen*). Accordingly, the present paper marks an attempt to shed light on our situatedness in language in terms of temporality. The investigation is thus concerned with the following question: what is the manner in which human beings arrive at and enter into language? That is to say, what kind of relation pertains between human existence and language when looked at from the perspective of temporality? Over the course of the investigation, the thoughts of other thinkers such as Heidegger, Derrida, Arendt and Merleau-Ponty will be considered as well.

I thus begin by first introducing the problem raised by Waldenfels against Gadamer, which concerns the question of radical foreignness set over against relative foreignness. I will then proceed to work out an interpretation of Gadamer's hermeneutics, which will consist in re-examining the ontological dimension of hermeneutics. I will take the notions of belonging (*Zugehörigkeit*) and hermeneutic distance (*Abstand*) as a decisive clue for this purpose. By elevating these notions to the ontological dimension, that is, to the dimension of language, it becomes possible to redraw the problem of foreignness brought forth by Waldenfels. In the process, my interpretation will place particular emphasis on Gadamer's appropriation of the notion of the

[1] Sloterdijk has actually written a book titled *Weltfremdheit*, but his aim in this book is not to thematise the phenomenon of foreignness as such, as in the manner of Waldenfels, but rather to examine the new way in which we relate to the world in contemporary society. He thus describes his contribution as a "phenomenology of worldless or world-estranged spirit" ("Phänomenologie des weltlosen oder weltabgewandten Geistes", Sloterdijk 1993, p. 13) (Translations are mine unless otherwise indicated.)

[2] I will explain the expression "non-difference" in section IV by drawing upon Merleau-Ponty's later thought.

"inpreconceivable"[3] (*Unvordenkliche*) in his later works that characterises our unique relation to the mother tongue, which is to be the central object of my investigation. Yet, insofar as we constantly need to make the mother tongue our home, the phenomenon of our coming-to-language (*Zur-Sprache-Kommen*) is already marked with a temporal displacement. The mother tongue is that which is bestowed upon us subsequent to our coming-to-the-world; and yet it is also at the same time that which we could not have had otherwise. It is here that Sloterdijk's analysis of the phenomenon of birth in relation to language gains relevance for our discussion. What therefore becomes crucial is the moment of our coming-to-language, the expression of which is used extensively by both Gadamer and Sloterdijk.[4]

2 The Demand of the Foreign

To respond to someone for whom the answer takes precedence over the question is already to have been questioned by that person. The fundamental criticism Waldenfels directs against Gadamer is brought out, among others, in his work *Vielstimmigkeit der Rede* (Polyphony of Speech).[5] Perhaps one can formulate his criticism as follows: hermeneutics is a way of thinking that is not capable of properly accounting for the phenomenon of foreignness. As he writes in relation to Gadamer, hermeneutic philosophy "concerns *relative foreignness for us*, not *foreignness in itself*".[6] Let us delve into this problem by considering it in a more proper context. What is at issue here is the expression "for us". According to Waldenfels, hermeneutics *relativises* the phenomenon of the foreign because it considers the phenomenon only within the bounds of the possibility of understanding. The foreign thus becomes that which is foreign *for us* and, as such, it turns into relative foreignness. Hermeneutics thus designates a process of appropriation (*Aneignung*) insofar as interpretation is a matter of making what is foreign familiar. Yet, as Waldenfels understands it, radical foreignness is precisely that which resists, escapes and lies beyond such a possibility of appropriation. Foreignness as such must not in any way be conceived in relation to us but what happens over and above all our interventions.

[3] The notion of the "*Unvordenkliche*" is usually rendered in English as either "immemorial" or "unprethinkable", but I shall prefer and use the expression "inpreconceivable" throughout the text. On the significance of this notion to Gadamer's later thought, refer to the following two articles by Jean Grondin (Grondin 1997 and Grondin 2000).

[4] While Gadamer's use of the expression "*Zur-Sprache-Kommen*" is conspicuous in the third part of *Truth and Method*, its significance is lost as the hyphens are omitted in the English translations.

[5] Waldenfels also offers a fairly elaborate critique of the hermeneutic model of dialogue in *Antwortregister* (Waldenfels 1994). While I have no room to discuss this issue here, he there brings into question the precedence hermeneutics attributes to the question over the answer. See chapter 13 of part one in the volume.

[6] "Es handelt sich lediglich um eine *relative Fremdheit für uns*, nicht um eine *Fremdheit in sich selbst*." (Waldenfels 1999, p. 71)

He therefore poses a rhetorical question: "Does the foreign let itself be dealt with within the ground of hermeneutics, or does it rather serve to call hermeneutics itself into question?"[7] His view is undoubtedly on the side of the latter.

Waldenfels' criticism is basically directed at the historical dimension of hermeneutics, since, as he sees it, the hermeneutic movement that is guided by tradition forms an immanent circle with itself. He thus interprets the aspect of hermeneutics which is determined as "participating in an event of transmission" (Gadamer 2004, p. 291; Gadamer 1990, p. 295.)[8] as indicative of a "will to meaning" (*Wille zum Sinn*, Waldenfels 1999, p. 72).[9] According to Waldenfels, hermeneutics therefore remains within Hegel's dialectic, insofar as the hermeneutic movement always departs from a particular tradition and returns to the very same tradition. The circle is thus closed and self-fulfilling:

> That this circle [of transmission] always opens itself *for us*, and that the horizon of meaning remains interminable and the meaning of a work or of a tradition remains inexhaustible – because the thing to be asked about and answered in the temporal diastase of prescription and recourse never corresponds to each other – do not eliminate the fact that the *circle itself* is already closed.[10]

Under this view, the phenomenon of the foreign will be accounted for only to the extent that it appears within a specific horizon of meaning. At the same time, however, one can recognise here the following fundamental presupposition lying behind Waldenfels' entire line of reasoning: the phenomenon of the foreign as such can only be properly grasped when it no longer resorts to the own.

One can certainly read Gadamer's *Truth and Method* as a work that attempts to establish a closed totality of transmission (*Überlieferung*) in its affirmation of the claims of tradition (*Tradition*), as many commentators have done including Waldenfels.[11] Yet, to begin with, it is worth mentioning that the work serves a propaedeutic and directive function in Gadamer's philosophy, as is indicated in its subtitle: "Fundamentals of Philosophical Hermeneutics" (*Grundzüge einer philosophischen Hermeneutik*).[12] Furthermore, the whole analysis in *Truth and*

---

[7] "Läßt sich das Fremde auf dem Boden der Hermeneutik bewältigen, oder ist dieses dazu angetan, die Hermeneutik selbst noch in Frage zu stellen?" (Waldenfels 1999, p. 67)

[8] Translation slightly modified.

[9] Needless to say, this expression has its recourse to Derrida, who had earlier accused Gadamer's hermeneutics of being committed to the Kantian ethic of the "good will". See Derrida 1989, p. 52.

[10] "Daß dieser Kreis [der Überlieferung] sich *für uns* immer wieder öffnet, daß der Sinnhorizont unabschließbar, der Sinn eines Werkes oder einer Überlieferung unausschöpfbar bleibt, weil Erfragtes und Geantwortetes sich in der zeitlichen Diastase von Vorgabe und Rückgriff niemals decken, schließt nicht aus, daß der Kreis *an sich* schon geschlossen ist." (Waldenfels 1994, p.136)

[11] For instance, see Bernasconi 1995, Caputo 1987 and Hamacher 1998.

[12] For some unknown reasons, this subtitle has been omitted in the English translations (both the translation by Barden and Cumin and that by Weinsheimer and Marshall). Jean Grondin writes the following in his essay "On the Sources of *Truth and Method*": "[O]ne should not limit TM [*Truth and Method*] to the work published in 1960. In fact, after 1960, Gadamer continued to work on TM, on his hermeneutics. Only the "Fundamentals" of TM remained, necessarily, the same. Those who wish to understand, to truly read TM, therefore, must also take into consideration both the works after and before TM. The composition of TM did not end in 1960 and, indeed, it still continues" (Grondin 1995, p. 98).

Method culminates, however incomplete, in the *ontological turn* that is carried out in the third part of the book. It is therefore in the realm of language that Gadamer's hermeneutics as a whole must be seen and assessed.[13] Thus it is insufficient, if not inappropriate, to characterise and deem his view as being reminiscent of Hegel's metaphysical totality simply on the grounds of his views on history and tradition.[14] As a matter of fact, insofar as Gadamer himself later comes to admit that "I was also acutely conscious of the fact […] that the third part of *Truth and Method* was only a sketch and I had not said everything I really had in mind",[15] one must take into account not only the third part of *Truth and Method* but the writings he produced throughout his long life in entirety.[16]

Needless to say, critics such as Waldenfels will not be convinced by general remarks such as these. But perhaps they can at least serve as a guide and motive for exploring the possibility of interpreting Gadamer's hermeneutics more faithfully by taking into account the manner in which he comes to develop his thought after the publication of *Truth and Method*. In the following section, I shall therefore attempt to reinterpret some of the key ideas in *Truth and Method* in light of Gadamer's later works, so as to elucidate the manner in which Gadamer's thought illuminates our diverse ways of being in language.

3 Beyond *Truth and Method* and Back

I wish to present a different interpretation of Gadamer's hermeneutics that diverges profoundly from Waldenfels' rendition of it. Here I shall attempt to twist and complicate the common narrative partly by reading Derrida through Gadamer and against Waldenfels. As a way into our subject matter, namely, the foreign, the mother tongue will be designated as our guide for the investigation. The choice for the

[13] Gadamer thus uses the expression "guidance" (*Leitfaden*) to describe his task in the third part of *Truth and Method*. He also subsequently adds the following footnote to *Truth and Method*: "Only in Part Three have I succeeded in broadening the issue to language and dialogue, though in fact I have had it constantly in view; and consequently, only there have I grasped in a fundamental way the notions of distance and otherness." (Gadamer 2004, p. 376; Gadamer 1990, pp. 316–317) Günter Figal also writes as follows in this regard: "Die Kunst und die hermeneutische Erfahrung der Geschichte werden in ihrem Sprachcharakter bestimmt, und es wird betont, daß sie allein in der Sprache möglich sind." (Figal 2007a, b, p. 219); Damir Barbarić also writes: "alle Überlegungen der beiden ersten Teile [der *Wahrheit und Methode* sind] eigentlich unterwegs zu einem ontologischen Ziel" (Barbarić 2007, p.199).

[14] Donatella Di Cesare responds to Waldenfels' critique in the following way: "It would be doing an injustice not only to Gadamer, however, but also to Schleiermacher to claim that the foreign would be *sublated*." (Di Cesare 2007, p. 211).

[15] "Freilich war mir dabei bewußt […] daß der 3. Abschnitt von *Wahrheit und Methode* nur eine Skizze war und nicht alles so sagte, wie ich es eigentlich im Auge hatte." (Gadamer 2007, p. 413; Gadamer 1997, p. 282).

[16] In this connection, it is noteworthy that the second volume of the collected works, which includes his writings from 1943 up until 1986, is also titled "Wahrheit und Methode", indicating the continuing development of his thought.

mother tongue is not arbitrary, for it offers a common ground upon which the differences between the views of Waldenfels and Gadamer can be best drawn out. For Gadamer, especially in his later writings, the mother tongue is closely associated with the notion of the home (*Heimat*). Following Heidegger, Gadamer comes to identify language as the place in which we dwell most fundamentally.[17] As such, Gadamer remarks in a late essay titled "*Heimat und Sprache*" (1992) that "Home is above all the home of language."[18] Yet, insofar as language is construed as the living language (*lebendige Sprache*) for Gadamer, this reference to home must not be understood as referring to rootedness (*Bodenständigkeit*, *Verwurzelung*) – as is the case with Heidegger, who invokes the original connection of the mother tongue to the earth. For, even if it may seem as though Gadamer is treading the same path as Heidegger when he expresses in another essay that "thinking occurs primarily in the mother tongue",[19] it must be made clear here that Gadamer introduces the idea of the mother tongue partly in order to differentiate his view from that of Heidegger's. While Heidegger also accords a privileged status to the mother tongue, he ultimately traces it back to our fundamental rootedness to the earth.[20] The human beings of the modern world are homeless (*heimatlos*), according to Heidegger, precisely because we have been uprooted from such an original ground and have lost our attachment to it.[21] In contradistinction to the poetic language of Heidegger, the essence of language for Gadamer lies not "under the earth", as it were, but in its very living character, as the living language we inherit and speak.[22] This is precisely why Gadamer can also make the claim that "One must always think in a language – although it does not always have to be the same language in which one thinks."[23]

Here I wish to highlight the manner in which Gadamer characterises our relation to language. According to Gadamer, the mother tongue marks a unique relation insofar as it "retains for every person something of inpreconceivable homeness".[24] By characterising our relation to the mother tongue as something inpreconceivable, the expression of which he appropriates from Schelling, Gadamer is bringing out

---

[17] For instance, see Heidegger's famous essay "Letter on Humanism" (Heidegger 1976).

[18] "Heimat ist vor allem Sprachheimat" (Gadamer 1993b, p. 366). As James Risser points out, this essay should be seen as Gadamer's response to Heidegger, insofar as he inverts the title of Heidegger's essay "Sprache und Heimat" (Heidegger 1983; see Risser 2012, p. 119.)

[19] "Aber es ist eine Aufgabe des Denkens, und Denken geschieht vor allem in der eigenen Muttersprache" (Gadamer 1993b, p. 428).

[20] In two essays that touch on the notion of *Heimat*, "Homecoming / To Kindred Ones" (Heidegger 1981) and "Letter on Humanism" (Heidegger 1976), Heidegger makes explicit connections of language to the earth (*Erde*) or ground (*Boden*).

[21] Among other essays by Heidegger that touch upon this subject, the essay "Letter on Humanism" (Heidegger 1976) is again very illustrative of this point.

[22] For the comparison of the notion of home in Heidegger and Gadamer, see the article "*Heimat* in Heidegger and Gadamer" by Kai Hammermeister (2000).

[23] "Man muß immer in einer Sprache denken – auch wenn es nicht immer die gleiche Sprache sein muß, in der man denkt." (Gadamer 1985, p. 353)

[24] "Die Muttersprache behält für jeden etwas von unvordenklicher Heimatlichkeit", (Gadamer 1993b, pp. 366–367).

the sense in which there is something wholly immemorial and inexplicable about the mother tongue.[25] In this sense, Gadamer's use of the notion "inpreconceivability" serves to distance and distinguish his own thought from that of Heidegger's.[26] Hence, in elucidating the phenomenon of the mother tongue, Gadamer does not define and characterise it as something merely trivial and self-transparent. On the contrary, he identifies intricate movements in our relation to the mother tongue such that it marks our paradoxical belonging to language. He thus writes:

> [The] magic [of the word the "inpreconceivable"] rests on the fact that we can perceive in it a real trace of this advance movement in our mind which always wants to think ahead and beyond; but over and over, again and again, it comes upon something that could never have been anticipated or planned for by using our imagination or by thinking ahead. That is *das Unvordenkliche*. Everyone knows something of this. […] I would remind you, for instance, of the [inpreconceivability] of home. […] One would like to uncover what is still shrouded in darkness, and yet one finds that it continually escapes us, and yet, for all that, it is always still there. (Gadamer 2007, p. 364; Gadamer 1985, p. 64)

Our relation to the mother tongue is distinguished by its being a place of both intimacy and estrangement. This idea is indeed reminiscent of the passage in *Truth and Method* where Gadamer remarks that "Language is so uncannily near our thinking" (Gadamer 2004, p. 370; Gadamer 1990, p. 383). Moreover, one also recalls in this connection a comment Arendt makes in a television interview about the special place the mother tongue occupies in her life in exile, where she responds: "What remains? The language remains." (Arendt 1994, p. 12; Arendt 1996, p. 59)[27] Shortly afterwards, she adds that "there is no substitution for the mother tongue." (Arendt 1994, p. 12; Arendt 1996, p. 59)[28] Yet, as Derrida reacts against Arendt with regards to these remarks, it must be admitted that a language can in fact go mad – the idea of which Arendt had explicitly denied in the interview. For a person or an entire people to go mad, such as was the case in Germany at the time of the war, the mother tongue must have already gone mad. Insofar as the mother tongue is that which is "irreplaceable", as Arendt suggests, it must also then be that which is responsible for our going madness. Focusing on the aspect of the *mother* of the mother tongue, this connection is also reflected in the word "uterus", which is derived from Latin and has its roots in the Greek word *hystéra* (ὑστέρα), thus suggesting that the entire process of pregnancy of the mother represents *hysteria*.[29] The "place of birth" of human beings is therefore marked by madness even from the

[25] See in particular *The Ages of the World* (Schelling 1946/2000), *Philosophie der Offenbarung* (Schelling 1977) and *Historical-Critical Introduction to the Philosophy of Mythology* (Schelling 1996/2007).

[26] Günter Figal provides a valuable insight in this connection in his book *Objectivity* (Figal 2010, pp. 9–17).

[27] "Was ist geblieben? Geblieben ist die Sprache." I here follow Donatella Di Cesare's analysis in her book *Utopia of Understanding* (Di Cesare 2012).

[28] "Es gibt keinen Ersatz für die Muttersprache."

[29] I absolutely do not intend on making this reference in a sexist manner. My point here is simply to point out an etymological and sociological point that what is called maternity was, etymologically speaking, already imbued with a form of madness called hysteria.

prenatal stage. This complex origin of our mother tongue is thus described by Derrida as the "most inalienable place" (*lieu le plus inaliénable*), a place we simply cannot abandon because the mother tongue always comes with us even in such an extreme case as being in exile. The mother tongue serves as a place of refuge in flight, a home we cannot let go even in migration. In this sense, the mother tongue is also the most mobile of all places precisely because it comes with us regardless of where we are. As Derrida points out, the mother tongue is therefore a "mobile home" (*chez-soi mobile*) as much as an "immobile home" (*chez-soi inamovible*) (Derrida 1997, p. 83; Derrida 2000, p. 89), a home that moves about with us in migration. The mobility of language is sustained precisely by its in-*alien*ability.

Such a characterisation seems appropriate for illustrating the sense in which Gadamer understands the mother tongue. For, having determined home as the home of language, he writes that home is the "place of original familiarity"[30] and that "home remains unforgotten".[31] It is a place we cannot simply choose to abandon and let go such as our place of dwelling, because the mother tongue always moves about with us wherever we may go. As Derrida puts it, "Language resists all mobilities *because* it moves about with me." (Derrida 2000, p. 91) Likewise, Gadamer suggests that "The horizon is, rather, something *into which* we move and that moves with us. Horizons change for a person who is moving." (Gadamer 2004, p. 303; Gadamer 1990, p. 309; emphasis added). What distinguishes Gadamer's account here is that we also move *into* a horizon. As we saw, the home for Gadamer is also something inpreconceivable, which suggests that our relation to the mother tongue can never be totalised so as to become complete and self-transparent. In this way, we can make sense of how Gadamer comes to assert that "Life is a sojourn in language",[32] for we truly enter into a language and make it our home, a home that is utterly irreplaceable.

I shall now attempt to tie Gadamer's understanding of the mother tongue back to the general framework of his hermeneutics. The sense in which the mother tongue is mobile is captured by the notion of hermeneutic distance. Instead of temporal distance, however, since what is at issue here is language, it is linguistic distance that is in question.[33] It is only within this realm of linguistic distance that we can properly speak of the mother tongue as one language among others. That is to say, the mother tongue, conceived within this realm, can be considered as one language among other foreign languages. The mother tongue as one language among others is thus countable, comparable and objectifiable. As we shall see, the question of foreignness arises only in the realm of distance. As Günter Figal thus writes in his

---

[30] "Aber was ist Heimat für uns, dieser Ort der Urvertrautheit?" (Gadamer 1993b, p. 367)

[31] "Die Heimat bleibt unvergessen", (Gadamer 1993b, p. 367)

[32] "Leben ist Einkehr in eine Sprache", Gadamer 1993b, p. 367).

[33] It is worth invoking here the footnote Gadamer later added to *Truth and Method* in the collected works edition: "it is distance, not temporal distance, that makes this hermeneutic problem [of distinguishing the true prejudices from the false ones] solvable." (Gadamer 2004, p. 376; Gadamer 1990, p. 304) See also another related footnote in Gadamer 2004, p. 376; Gadamer 1990, pp. 316–317.

essay "Fremdheit und Entferntheit": "What is interpretable and demands to be interpreted is not what is foreign, but what is distanced."[34] In contrast, the sense of the immobility of the mother tongue, the second aspect of the mother tongue, will be captured by Gadamer's notion of belonging. Since the language to which we belong marks our home, the place of original familiarity, it cannot be reduced to being a mere language, one language among others. In this sense, the mother tongue is not comparable to other languages, since it is precisely that which is in essence irreplaceable and irreducible. It therefore has the character of *verticality*, which will be of importance when we discuss the notion of non-difference as elucidated by Merleau-Ponty. For the moment, however, I shall attempt to outline in the next section the implications of the analysis carried out above, and how it challenges the general understanding of the foreign in Waldenfels.

4 The Mother of Language

One immediate consequence of the analysis carried out above is the recognition that the question of the foreign is not at all primary. For it is only within the dimension of linguistic distance that we can speak of a *foreign* language at all, in contradistinction to the mother tongue. On the other hand, Gadamer's hermeneutics shows that the mother tongue figures above all as something that is absolutely singular and unique in our relation to language. The manner in which Waldenfels picks out the foreign as that which is anywhere and at all times absolutely primary must thus be questioned. Up until now, however, I have worked out the phenomenon of the mother tongue only from the perspective of Gadamer's hermeneutics. Let us now see how Waldenfels understands the mother tongue. We shall see that it is not only through the hermeneutic critique that his view can be seen to be problematic, but his view as such will be exposed as deeply misguided and erroneous.

In an interview, Waldenfels makes the following observation, which may at first sight seem quite trivial and innocent:

> The own originates from the contrast with the foreign. If everyone spoke the same language, then I would have no mother tongue. I would simply speak. Every child discovers a language through the language of their parents, which for them is initially a foreign language.[35]

Upon a closer observation, however, we should come to realise that his remark is self-contradictory. On the one hand, it implies that the child would be speaking a "foreign" language, even in the case where everyone spoke the same language – for whatever the language a child learns from their parents is supposed to be "foreign".

[34] "Interpretierbar und interpretationsbedürftig ist nicht das Fremde, sondern das Entfernte." (Figal 2009, p. 227)

[35] "Das Eigene entspringt dem Kontrast mit dem Fremden. Wenn alle die gleiche Sprache sprächen, so hätte ich keine Muttersprache. Ich würde einfach sprechen. Jedes Kind entdeckt die Sprache durch die Sprache der Eltern, die für ihn zunächst eine Fremdsprache ist." (Rotaru 2010, p. 255)

On the other hand, there cannot be a foreign language, at least according to this view, when everyone spoke the same language. His remark thus contradicts itself. That is to say, Waldenfels entirely overlooks the fundamental character of language whereby we first and foremost enter into it and make it our home. The mother tongue marks precisely a language that is made one's own prior to the state in which the difference of the own and the foreign arises. Accordingly, there is no discrimination between the own and the foreign in the language acquisition of children. It would be quite arrogant of a view if adults simply assume that the same sort of relation to language holds for children as it does for them when they learn a *foreign* language.[36] I shall briefly come back to this discussion in the conclusion in order to draw out the significance of these ideas for such disciplines as linguistics and developmental psychology.

Waldenfels does offer insights elsewhere which are perhaps a little more coherent, yet his view remains at bottom problematic. Let us take a look at two other passages where he discusses the question of the mother tongue: "The contrast between the own and the foreign applies even for the learning of the mother tongue in children. It goes hand in hand with the becoming foreign of other languages."[37] In another text, he writes: "The mother tongue enjoys privilege not because it is better than other languages, but because it is that language in which one is introduced to the realm of language. Foreign languages originate even then, when one learns his or her own language."[38] While his view is developed perhaps a little more subtly here, he still maintains the idea that the phenomenon of the foreign can never be dissociated from the own. There is nothing primary about the mother tongue insofar as the foreignness of language always accompanies our relationship to language. In other words, he employs what I would call the *logic of difference*, where the own and the foreign are at all times set against each other.[39] But such a view fundamentally fails to take into consideration the respect in which the mother tongue is singular and irreplaceable. If the mother tongue can be understood as that to which we

[36] It is worth noting here that Habermas, in his review of *Truth and Method*, locates in Gadamer's hermeneutics an important distinction between learning of a language (*eine Sprache lernen*) and learning to speak (*Sprechenlernen*), the latter of which is of course at work in the case of children (See Habermas 1988, p. 145). Wilhelm von Humboldt had already made this distinction in his *On Language* (See von Humboldt 1998, chapter 9, in particular pp. 58–64).

[37] "Es gibt ebenso für das Erlernen der Muttersprache, das mit einem Fremdwerden anderer Sprachen Hand in Hand geht" (Waldenfels 1997, p. 156).

[38] "Die Muttersprache genießt nicht etwa deshalb einen Vorzug, weil sie besser ist als andere Sprachen, sondern weil sie jene Sprache ist, in der jemand in das Reich der Sprache eingeführt wurde. Fremdsprachen entstehen eben dann, wenn man die eigene Sprache erlernt. Diese läßt sich ebenso wie jene *als eine Sprache unter anderen* betrachten, und wir tun dies, wenn wir Sprach- oder Kulturvergleiche anstellen. Doch die Betrachtung von Sprache oder Kultur als einer Sprache oder einer Kultur unter anderen folgt einem Gesichtspunkt, der die entsprechende Differenz von Eigenem und Fremdem nicht generiert, sondern supponiert und das, was in actu unvergleichlich ist, einem Vergleich unterzieht" (Waldenfels 1999, p. 179).

[39] Di Cesare uses the expression "thought of difference" and places Waldenfels' view in question in a similar way: "The charge from those who want to give priority to the foreign as a radical provocation is, therefore, unwarranted" (Di Cesare 2012, p. 212).

belong at the most intimate dimension, it is precisely because the mother tongue is not one language among others; it is language *par excellence*. Derrida therefore identifies the mother tongue as "the first and the last condition of belonging" (Derrida 2000, p. 88), the absolute condition of belonging that both appropriates and expropriates. With regards to his own testimonial remark that "the only language I speak is not mine", Derrida makes the following point:

> I draw your attention to a first slippage: up until now, I have never spoken of a "foreign language." When I said that the only language I speak is *not mine*, I did not say it was *foreign* to me. There is a difference. It is not entirely the same thing (Derrida 1998, p. 5, emphasis added).

It is these differences that are essentially neglected by Waldenfels due to the logic of difference on which his view rests.[40] In contrast, Gadamer's hermeneutics offers an intricate understanding of our relation to the mother tongue that not only encapsulates the sense of linguistic distance but also the sense of our unique and irreplaceable belonging to it. Gadamer thus writes as follows:

> However thoroughly one may adopt a foreign frame of mind, one still does not forget one's worldview and language-view. Rather, the other world we encounter is not only foreign but is also related to us. It has not only its own truth *in itself* but also its own truth *for us*. (Gadamer 2004, p. 439; Gadamer 1990, p. 445)

I have tried to show such paradoxical characters of our relation to the mother tongue particularly by attending to the notion of the inpreconceivability of the home which Gadamer develops in his later writings. Since our analysis has hitherto remained mostly descriptive, however, I now wish to provide some theoretical grounding for these linguistic characters of Gadamer's hermeneutics by identifying a moment of non-difference, which is to be brought out by examining the phenomenon of birth explored by Sloterdijk.

5 Forgetfulness of Birth

Waldenfels' understanding of the mother tongue is illustrative of the problem inherent in the thinking of difference. Hence our analysis has reached the point where we now raise the question concerning the moment of non-difference, the moment at which the notion of the inpreconceivable in Gadamer is to be located. In order to properly ground the privileged place occupied by the mother tongue in this sense, we must now turn to the examination of its temporal character. Our investigation is to be guided by the two prefixes in the expression "in-pre-conceivable" (*un-vordenklich*), which indicate its temporal characteristics in its "not" (of the prefix "in") and "before" (of the prefix "pre"). It is to be determined and understood above all as

[40] Although I have no space to work out the thesis here in this paper, I essentially make a distinction between the *mother* tongue (*Muttersprache*) and the *own* language (*Eigensprache*). This distinction will be further investigated in my on-going research.

that which we cannot get behind and make it an object of thought.[41] The question that needs to be raised and answered is therefore the following: on what grounds is the mother tongue to be designated and legitimised as that which is we cannot get beyond? Furthermore, how are we able to examine and describe that which we cannot conceive by getting beyond? To this end, Sloterdijk's analysis of birth and language can offer some valuable insights.

The work by Sloterdijk that is of most importance to us is *Zur Welt kommen – Zur Sprache kommen*, which is a book that is based on a series of lectures he held at Goethe University in Frankfurt in 1988. There he identifies various factors that determine our finite relation to the world, above all, as the title suggests, with respect to the manner in which our coming-to-the-world must be thought concurrently with our coming-to-language. Yet this is not meant to imply in any way that our relation to the world will thereby be reducible to that of language. In truth, Sloterdijk identifies and thematises in this work a temporal displacement that occurs between the moment of birth and the moment of entering into a language. On the one hand, it is undeniable that we are brought over to the world without being tied to a specific language a priori. For we learn to speak, in most instances in any case, the language of the "mother" that has been bestowed upon us subsequent to our coming-to-the-world (aposteriority of language). On the other hand, however, it is just as undeniable that we can hardly speak of a "world" set apart from our relation to language. For, as we will see, we are utterly unable to understand our own being outside of language, even if we would of course, in theory, still be "in space" as a physically occupying body regardless of such a factical condition (apriority of language). Precisely in this sense, Gadamer remarks that "In truth we are always already at home in language, just as much as we are in the world." (Gadamer 1976, p. 63; Gadamer 1993a, p. 149) Accordingly, we are also led to acknowledge the a priori dimension of language which Gadamer calls "linguisticality" (*Sprachlichkeit*). There thus arises an important question as to what extent we are "in the world" prior to acquainting ourselves with a language. Let us at this point approach this problem from a slightly different point of view.

Inpreconceivability designates a moment of non-difference insofar as the mother tongue marks that which we cannot get beyond, thus does not operate within the logic of difference between the own and the foreign. The expression *non-difference* is here to be understood in the sense of original belonging that is to be distinguished from mere identity and difference. In this connection, it is worth evoking the description Merleau-Ponty offers in his working notes in *The Visible and the Invisible*:

> The *antecedent* unity me-world, world and its parts, parts of my body, a unity before segregation, before the multiple dimensions – and so also the unity of time – – Not an architec-

[41] Interestingly, Gadamer and Sloterdijk employ similar expressions to convey this point. Gadamer indicates that the inpreconceivable is to be understood as that which we cannot get behind (*kann nicht mehr dahinterkommen*), whereas Sloterdijk conceives of the mother tongue as the condition which we cannot get beyond (*unhintergehbare Bedingungen*) (See Gadamer 2007, p. 364; Gadamer 1985, p. 64, and Sloterdijk 1988, p. 162).

> ture of noeses-noemata, posed upon one another, relativizing one another without succeeding in unifying themselves: but there is first their underlying bond by *non-difference* -- All this is *exhibited* in: the sensible, the visible.[42]

In this sense, non-difference is here to be understood as "antecedent unity" and "underlying bond" that can be thought of as neither difference nor identity – what Merleau-Ponty also calls "wild being" and "vertical world" in the above work. Gadamer's sense of belonging as revealed in our relation to the mother tongue is here picked up by the notion of non-difference. Moreover, it is to this pre-predicative sphere of non-difference that Sloterdijk directs his attention in his analysis of birth and language. By alluding to the story *The Book of Sand* by Jorge Luis Borges, Sloterdijk underscores the significance of the factical condition whereby the story of our life has already begun to be written before we ourselves have become the author of it. Like the surreal book depicted in Borges' story, whose beginning cannot be opened no matter how many pages one turns, our life had already begun before we ourselves became aware of it.[43] In most cases, we come to find out about our own birth – for instance, how, where and when we came into the world – through the indirect accounts given to us by our birth parents and other possible witnesses. None of us were at our own beginnings ourselves. As Sloterdijk writes, it is as if we came "too late into the theater"[44] and have missed the beginning of the play. Our only choice is therefore to try to follow the rest of the play the best we can. This illustration indeed applies just as well to our relation to language. For we come to speak the mother tongue prior to becoming able to write the life story ourselves, or prior to becoming able to choose to write in another language, since the language of our life story is already determined at infancy.

Yet Sloterdijk's view is distinguished from the thinking of difference by the fact that he does not do away with our finitude by employing such concepts as the Foreign and the Other. The human condition is marked not merely by such a dimension of displacement but also by the dimension of absolute singularity. For our birth is undoubtedly our *own* birth, and can never manifest as that of someone else's, regardless of whether the actual experience of our coming-to-the-world is available to our memory. Despite our inability to retain the memory of our coming-to-the-world and coming-to-language, we are permanently impressed by birthmarks (including the umbilicus) and what Sloterdijk calls a "linguistic tattoo"[45] (*linguistische Tätowierung*). Accordingly, Sloterdijk argues in another work that birth is

---

[42] Merleau-Ponty (1968), p. 261. There is undoubtedly an interesting connection here to Kant, Schelling and Husserl in this regard, all of whom characterise the origin (*Ursprung*) as indifference (*Indifferenz*).

[43] In the actual story, it is both the beginning and the end of the book which we can never reach. With regards to our life, however, it is primarily our birth that lies beyond the scope of our memory, insofar as our death is something we can technically get quite close to. This is exemplified, for instance, by the expression "near-death experience", whereas we typically do not speak of "near-birth experience".

[44] "Zeitlebens [...] sind wir in der Lage von Leuten, die zu spät ins Theater kommen", (Sloterdijk 1988, p. 12).

[45] Sloterdijk 1988, p. 157.

much more fundamentally *mine* than death is in Heidegger's sense of mineness (*Jemeinigkeit*).[46] As Sloterdijk puts it, "the immemoriality of my birth is – far more than the Heideggerian anticipation of one's own death – my existentiell signature".[47] In Sloterdijk's view, our factical condition exemplified by birth is undeniably singular and absolute, even if it can never appear transparent to us. For this reason, he describes this state of opaqueness as "my enabling *darkness*" (*eine mich ermöglichende Dunkelheit,* Sloterdijk 1993, p. 238, emphasis added). As such, we can speak of inherent belonging that binds us to our own existence, which we are calling here "non-difference". In this respect, while Sloterdijk might appear to be echoing Arendt's point concerning the primacy of natality over mortality, one must draw a fine distinction here insofar as the latter claims that we "are born into the world as strangers." (Arendt 1998, p. 9) Since the non-differential belonging precedes any difference between the familiar and the strange, our birth cannot be characterised in terms of strangeness or foreignness. Indeed, as I have attempted to work out above, it is precisely in this sense that Gadamer's use of the notion the "inpreconceivable" is to be understood. As was already quoted, he writes with respect to the inpreconceivable that "One would like to uncover what is still shrouded in *darkness*, and yet one finds that it continually escapes us, and yet, for all that, it is always still there." (Gadamer 2007, p. 364; Gadamer 1985, p. 64, emphasis added). It is particularly noteworthy that the word "darkness" (*Dunkeln*) is used here by Gadamer as well. Insofar as original belonging marks that which is "always still there" and yet at the same time that which "continually escapes us", it concerns the ontological realm that essentially precedes the logic of difference. For there can strictly be no opposition in this realm between my birth and that of others, nor between my mother tongue and other languages.

The temporality that pertains to the original belonging of our coming-to-the-world and coming-to-language can be elucidated in terms of forgetfulness. Just as Sloterdijk puts forward the expression "forgetfulness of birth" (*Geburtsvergessenheit*, Sloterdijk 1988, p. 60; Sloterdijk 1993, p. 237), which I have just worked out above, Gadamer also invokes the expression "forgetfulness of language" (*Sprachvergessenheit*, Gadamer 2004, p. 418; Gadamer 1990, p. 422) in *Truth and Method.* Needless to say, Gadamer introduces this expression with respect to the manner in which the concept of language has been objectified over the course of the history of Western thought – specifically with regards to the concept of the word that has, since Plato's *Cratylus*, constantly been misinterpreted as mere sign or name – it is nonetheless fitting to draw a connection between this expression and our current discussion concerning the mother tongue.[48] For the mother tongue is, partly due to the logic of difference, just as objectified and instrumentalised, insofar as it is

[46] Concerning Heidegger's concept of mineness, refer to sections 47 and 51 of *Being and Time*.

[47] "Meine Nicht-Erinnerung an meine Geburt ist – viel mehr als das Heideggersche Vorlaufen in den eigenen Tod – meine existentielle Signatur." (Sloterdijk 1993, p. 239)

[48] Gadamer writes as follows: "This [replacement of the concept of the image by that of the sign] is not just a terminological change; it expresses an epoch-making decision about thought concerning language." (Gadamer 2004, p. 414; Gadamer 1990, p. 418).

thought of as merely one language among others. Our coming-to-the-world and coming-to-language are thus inpreconceivable because these two moments are, on the one hand, already forgotten, and yet, on the other hand, never ceases to be there for us. These theses will be further developed in the following section.

6 Transmission and Relief

Let us at this point briefly touch on some of the issues surrounding the relation between birth and language in order to highlight some possible tensions between the thoughts of Gadamer and Sloterdijk. While it is true that the two notions mark a moment of non-difference, they are undeniably two distinct moments as well. One is first born; only afterwards does one acquaint oneself with a language. In this connection, Sloterdijk remarks that "This [fictional book in Borges' story] says nothing else than the fact that for human beings, as finite speaking beings, the beginning of being and the beginning of language do not coincide."[49] Moreover, if language is to be universal and to serve as the mode by which the tradition is transmitted, as Gadamer has so convincingly demonstrated, there arises an important question as to what extent, if at all, we are capable of individuating ourselves from our belonging to it. For this is precisely the point at which Sloterdijk seeks to diverge most explicitly from Gadamer's view. Whereas Gadamer has rightly recognised the aspect in which language is "handed down" (*überliefert*, *weitergegeben*) to us, it is just as necessary, according to Sloterdijk, to account for the possibility of the language becoming *our own* language, and not simply remain the language of the "mother". This critique must not be confused with those which have been raised against Gadamer, for instance, by Ricœur and Habermas concerning the problems of distanciation and emancipation respectively. For at issue here is the very modality of the transmission of language and it does not involve casting doubt on the effect of history or the authority of tradition themselves, as it does in the case of the critiques by Ricœur and Habermas.[50] In this sense, Sloterdijk's critique must be taken very seriously, insofar as it is an *immanent* critique that is raised within the framework of Gadamer's hermeneutics itself. Sloterdijk indicates that "the language, which is

[49] "Das besagt nichts anderes, als daß für Menschen, als endliche sprechende Wesen, der Seinsanfang und der Sprachanfang unter keinen Umständen zusammenfallen. Denn fängt die Sprache an, so ist das Sein schon da; will man mit dem Sein beginnen, versinkt man im schwarzen Loch der Sprachlosigkeit." (Sloterdijk 1988, p. 38)

[50] In fact, the opposite is the case here. Whereas the question posed by Ricœur and Habermas remained a matter of drawing distance from and critically reflecting on that which has been transmitted to us – so as to gain a more critical and objective ground on our own situation – the question here developed by Sloterdijk concerns the possibility of appropriating the tradition in its utmost intimacy so that the process of appropriation achieves its true singularity and originality. I shall make reference to the process of *cultivation* or *formation* (*Bildung*) in Gadamer. Precisely by cultivating the tradition as our own, the tradition is no longer something that stands at a distance but rather becomes that which we above all think and speak.

being directly handed down to us, is already the language of our political community at birth."[51] Hence, according to Sloterdijk, the language that is handed down to us is in fact a national language (*Nationalsprache*), a language that is already institutionalised and politicised. He therefore highlights the need to theoretically account for the possibility of internalising such a form of language so as to bridge the distance opened up by it, insofar as it is ideology-laden and not individuated as *our own* language.

This political dimension of language harks back the discussion of Derrida, whose enigmatic remark that "the only language I speak is not mine" (*Je n'ai qu'une langue, ce n'est pas la mien,* Derrida (1998/1996), p. 1) actually carries a political tone. For this testimonial remark exhibits precisely the impossibility of identifying the national language as *our own* language. This is what Derrida describes with the expression "monolingualism of the Other": the language that is singular but does not and can never belong to me as my own language. Insofar as Derrida speaks of and insists on the impossibility of making the singular language our own, however, we must deem the manner in which Derrida formulates his thesis "negative". This can be shown above all by considering the following passage:

> The language called maternal is never purely natural, nor proper, nor inhabitable. *To inhabit*: this is a value that is quite *disconcerting* and equivocal; one never inhabits what one is in the habit of calling inhabiting. There is no possible habitat without the difference of this exile and this nostalgia. Most certainly. That is all too well known. But it does not follow that all exiles are equivalent. From this shore, yes, *from this* shore or this common drift, all expatriations remain singular. (Derrida 1998/1996, p. 58)

While Derrida fully acknowledges the singularity of the mother tongue, his aim consists in undermining its proper place by displacing its foundations – what he calls "expatriation". In this respect, we can claim that Derrida slips the thinking of difference into the realm of non-difference. In contrast, Sloterdijk makes a fine distinction between the two poles by treading the path that is "otherly positive"[52] (*das Andere des Positiven*). Even if there is only a subtle difference between Sloterdijk and Derrida, it is undoubtedly a significant subtlety of difference. For, in speaking of the positivity of our relation to language, Sloterdijk does not deny and object to the basic assumption of the hermeneutical thinking, that we always find ourselves situated in a factical condition.

Sloterdijk locates a possibility of taking a critical stand on the mother tongue in what he calls "releasement" (*Entbindung*). Just as we were detached (*entbunden*) from our mother upon birth, it is also necessary for us to release ourselves from the constant mediation of the national language. In the case of language, such a moment is brought about in the form of a relief (*Freispruch*[53]) of the language. That is to say,

---

[51] "Die Sprache, die uns auf dem Weg der unmittelbaren Weitergabe nahegegangen ist, ist immer schon die Sprache unserer politischen Geburtsgemeinschaft." (Sloterdijk 1988, p. 154)

[52] Sloterdijk 1988, p. 164. He also writes as follows: "Die anderen Wege, zu einem sogenannten kritischen Bewußtsein zu kommen, sind selbst schon positiv." (ibid., p. 164)

[53] Sloterdijk certainly has in mind here the literal sense of this German word: "to speak freely" (*frei sprechen*).

the mother tongue is relieved from its national authority at the moment where the moment of our birth is brought back in view in and by the language. In this sense, the essence of language lies in its ability to bring back the beginning as a new beginning, in the poetics of the beginning. Within the transmission of the language, one is called upon to find one's own language that evokes the moment of birth. However, insofar as our beginning, as an event that precedes our memory, remains in darkness for us, our language must illuminate and bring out that which is unknowable, that which cannot be assimilated in thought. Sloterdijk thus writes as follows:

> To language […] belongs the breath of relief. This releases us from the organic nationality and the fallenness into ensuing violence. It calls back the first moments of being-in-the-world, in which the experience of air in every touch of outside carries along the maternal element. The breath, from the beginning, as the first instance of natality, goes beyond nature with nature.[54]

For Sloterdijk, the essence of language consists in preserving the very dimension which is hidden from our view but which profoundly shapes and defines us. Only in this way can a language become that of *our own*, the *way* of our *tongue* proper (*Mundart*).[55]

With respect to the standpoint of Gadamer's hermeneutics in this regard, one can meet Sloterdijk's challenge by pointing out that, precisely for this reason, Gadamer refrains from politicising his views on language at the most general level. For that would involve positing a counter-movement that allows for critical individuation.[56] Rather than emphasising the need to seek "another" language that would be free of any (ideological) distortion, Gadamer sought to bring to light the infinite possibility that inheres in the phenomenon of language itself. One aspect in which such a force of language can be brought out is by attending to the idea of the word that occupies a central place in Gadamer's hermeneutics, and this can help address, even if in a limited fashion, the problem posed by Sloterdijk in a more concrete manner.

If there is a risk inherent in the mere transmission of a language, such a risk is to be found in our conception of language itself. As Gadamer works out in the third

---

[54] "Zu der Sprache, die den Zuruf zwischen Zurweltgekommenen artikuliert, gehört auch der Atem des Freispruchs. Dieser entbindet uns von der naturwüchsigen Nationalität und von der Verfallenheit an die erworbene Gewalt. Er ruft die ersten Momente des In-der-Welt-Seins zurück, in denen die Lufterfahrung jeder äußeren Berührung mit dem mütterlichen Element vorangeht. Der Atem als erste Instanz der Natalität geht von Anfang an mit der Natur über die Natur hinaus." (Sloterdijk 1988, pp. 165–166)

[55] Here my interpretation of Gadamer actually comes closer to Heidegger's view of language as exemplified by such essays as "Sprache und Heimat" by the latter. It is still my view that Heidegger is mistaken in tracing language back to the rootedness to the earth in speaking of the dialect (*Mundart*) – an important distinction that must be maintained between the views of the two thinkers.

[56] Precisely by not wholly endorsing the movement of individuation, I believe Gadamer's view of language is actually better able to identify the authenticity of the mother tongue through community and the collective. In other words, language as a dialect or a mother tongue is not merely to be understood in the sense of a language of the individual, but also as that which has already brought us in relation with others. This subject matter, however, is outside the scope of this paper and will need to be further worked out elsewhere.

part of *Truth and Method*, there has historically been a "forgetfulness of language" where language has been understood as a system of signs rather than as the accomplishment of the word we bring to speech. What Gadamer calls the "valence of being" (*Seinsvalenz*, Gadamer 2007, p. 152; Gadamer 1993b, p. 54) marks the manner in which the inner infinity, the whole of being, comes into language precisely as a word. Only in this sense does he speak of the "truth of the word", and such an event is accomplished above all in the words of poetry:

> The universal "there" of being that resides in the word is the miracle of language, and the highest possibility of saying consists in catching its passing away and escaping and in making firm its nearness to being. It is nearness or presentness not of this or that but of the possibility of everything. This is what distinguishes the poetic word. (Gadamer 2007, p. 152; Gadamer 1993b, pp. 54-55)

In this sense, Gadamer addresses another sense of "relief" and "detachment" that is, like Sloterdijk, brought about in and through poetry. As such, his concern was not to provide an account of how we can take a critical stand on the language we inherit, but to recall how truth is manifested in the saying of the word. Hence, just as Gadamer evokes the doctrine of aesthetic non-differentiation (*ästhetische Nichtunterscheidung,* Gadamer 2004, p. 116; Gadamer 1990, p. 122) in order to challenge and contest the qualitative distinction made between the picture (*Bild*, the image) and its original reality (*Urbild*, the source), there is also a process of *linguistic* non-differentiation at work here in the word. What he calls the "valence of being of the picture" in *Truth and Method* is here worked out as the valance of being of the word.[57] The crucial distinction with the linguistic non-differentiation from the aesthetic, however, is that the saying of a word reverberates the whole of language by its being related to the unsaid – a characteristic that is missing from the representation of a picture (Gadamer 2007, p. 152; Gadamer 1993b, pp. 54–55). The idea of *valence* therefore carries as its essential character the sense of *non-difference*. The truth of the word is revealed in a word which brings out this very non-difference, and this is accomplished above all in the poetic word. The language we speak thus retains its truth, even if our first encounter with it remains in darkness for us, insofar as the poetic word crystallises and re-expresses such bygone moments. In this way, Gadamer's conception of the word also brings to our attention, even if in a different manner from Sloterdijk, that the poetic word, understood within the subject matter under discussion here, brings out the original moment of birth by constantly illuminating the non-difference of the word and being. Lastly, I believe it is important to bring to attention that, insofar as Sloterdijk also indicates that "the way into the open runs through the middle of language itself",[58] the crux of his thought can generally be seen to be in line with Gadamer's view.

[57] Lawrence K. Schmidt has written a concise commentary on this development in Gadamer's thinking (Schmidt 1995).

[58] "[D]ie Spur ins Freie läuft mitten durch die Sprache selbst" (Sloterdijk 1988, p. 165).

7 Conclusion

On the whole, I believe the philosophical hermeneutics as worked out by Gadamer addresses, or is capable of addressing, many of the important problems that have been raised by Waldenfels and Sloterdijk and, in this regard, his thought still remains relevant to and contemporary with us. In this paper, I have brought forth an interpretation of Gadamer's hermeneutics, even if in an all too limited manner, in order to advance the thesis that hermeneutics must be thought beyond the dichotomy between radical and relative foreignness by rejecting the logic of difference. By considering Gadamer's views on language in light of Sloterdijk's works, I have attempted to highlight in particular the manner in which the coming-to-language marks a moment of non-difference that is displayed in our relation to the mother tongue. In ordinary terms, this insight is reflected in the fact that every child is, theoretically speaking, not only capable of learning every human language in existence, but is also capable of making it his or her home as the mother tongue.[59] No infant is born tied to a particular language a priori, as if all other languages would then somehow manifest as foreign.[60] Had that been the case, Derrida's "monolingualism" would not have arisen as a fundamental problem of language and politics. Insofar as the question of learning to speak in infants is taken up in philosophical terms, this contribution serves to bring philosophical insights into dialogue with the work done in such fields as linguistics, developmental psychology and cognitive science. I hope this investigation contributes, by way of non-difference, to demonstrating the universality of hermeneutics, which remained at the heart of Gadamer's hermeneutical thinking throughout.

[59] Those who grow up with more than one language (simultaneous multilingualism) may initially undergo a phase where language differentiation has not yet been made explicitly (See, for instance, Genesee and Nicoladis 2007; Du 2010, and Meisel 2006). One can interpret this phenomenon as having to do with the fact that such multilingual infants are learning to *speak* and not (merely) learning *a language*. In other words, it is precisely because each so-called "language" is not experienced as a foreign language that the infants can relate to the languages spoken around them in an unmediated, intimate way, thereby giving rise to such occurrences as code-switching, language mixing and so on.

[60] Interestingly, Sloterdijk makes the following remark with regards to nationality in an interview: "A newborn is not yet a Swiss, not yet a German, not a Chinese. Because we are where we are, we will also be locally socialised." ("Ein Neugeborenes ist noch nicht Schweizer, noch nicht Deutscher, nicht Chinese. Weil sie sind, wo sie sind, werden sie auch lokal sozialisiert", Sloterdijk 2013, p. 71).

References

Arendt, H. 1994. *Essays in understanding*, 1930–1954. New York: Harcourt Brace.

———. 1996. *Ich will verstehen: Selbstauskünfte zu Leben und Werk*. München: Piper.

———. 1998. *The Human Condition*. 2nd ed. Chicago: University of Chicago Press.

Barbarić, D. 2007. Die Grenze zum Unsagbaren: Sprache als Horizont einer hermeneutischen Ontologie (GW 1, 442–478). In *Hans-Georg Gadamer: Wahrheit und Methode. Klassiker Auslegen, Band*, ed. G. Figal, vol. 30, 199–218. Berlin: Akademie Verlag.

Bernasconi, R. 1995. "You Don't know what You're talking about": Alterity and the hermeneutic ideal. In *The Specter of Relativism: Truth, Dialogue, and Phronesis in Philosophical Hermeneutics*, ed. L.K. Schmidt, 178–194. Evanston: Northwestern University Press.

Caputo, J.D. 1987. *Radical Hermeneutics: Repetition, Deconstruction, and the Hermeneutic Project*. Bloomington/Indianapolis: Indiana UP.

Derrida, J. 1989. Three questions to Hans-Georg Gadamer. In *Dialogue and Deconstruction: The Gadamer-Derrida Encounter*, ed. D.P. Michelfelder and R.E. Palmer, 52–54. Albany: State University of New York.

———. 1997. *De l'hospitalité*. Paris: Calmann-Lévy. English edition: Derrida, J., & Dufourmantelle, A. (2000). *Of Hospitality*. Trans. R. Bowlby. Stanford: Stanford University Press.

———. 1998. *Monolingualism of the Other: Or, The Prosthesis of Origin*. Trans. P. Mensah. Stanford: Stanford University Press. French edition: Derrida, J. 1996. *Le monolinguisme de l'autre: ou la prothèse d'origine*. Paris: Galilée.

Di Cesare, D. 2007. *Gadamer: A Philosophical Portrait*. Trans. N. Keane. Bloomington: Indiana University Press.

———. 2012. *Utopia of understanding: Between babel and Auschwitz*. Trans. N. Keane. Albany: State University of New York.

Du, L. 2010. Initial bilingual development: One language or two? *Asian Social Science* 6 (5): 132–139.

Figal, G., ed. 2007a. *Hans-Georg Gadamer: Wahrheit und Methode. Klassiker Auslegen, Band 30*. Berlin: Akademie Verlag.

———. 2007b. *Wahrheit und Methode* als ontologischer Entwurf: Der universale Aspekt der Hermeneutik (GW 1, 478-494). In *Hans-Georg Gadamer: Wahrheit und Methode. Klassiker Auslegen, Band*, ed. G. Figal, vol. 30, 219–236. Berlin: Akademie Verlag.

———. 2009. *Verstehensfragen: Studien zur phänomenologisch-hermeneutischen Philosophie*. Tübingen: Mohr Siebeck.

———. 2010. *Objectivity: The Hermeneutical and Philosophy*. Trans. T.D. George. Albany: State University of New York.

Gadamer, H.-G. 1976. Man and language. In H.-G. Gadamer, *Philosophical Hermeneutics* (pp. 59–68), Trans. and ed. D. E. Linge. Berkeley: University of California Press.

———. 1977. *Philosophical Hermeneutics*. Trans. D. E. Linge. Berkeley: University of California Press.

———. 1985. *Gesammelte Werke*, Hermeneutik im Rückblick. Vol. 10. Tübingen: Mohr Siebeck.

———. 1990. *Gesammelte Werke*, Hermeneutik I: Wahrheit und Methode. Grundzüge einer philosophischen Hermeneutik. Vol. 1. Tübingen: Mohr Siebeck.

———. 1993a. *Gesammelte Werke*, Wahrheit und Methode: Ergänzungen Register. Vol. 2. Tübingen: Mohr Siebeck.

———. 1993b. *Gesammelte Werke*, Ästhetik und Poetik. Vol. 8. Tübingen: Mohr Siebeck.

———. 1997. *Gadamer-Lesebuch*, ed. J. Grondin. Tübingen: Mohr Siebeck.

———. 2004. *Truth and Method*, Rev. ed., Trans. J. Weinsheimer, and D. G. Marshall. London: Continuum.

———. 2007. *The Gadamer Reader: A Bouquet of the Later Writings*. Trans. R.E. Palmer. Evanston: Northwestern University Press.

Genesee, F., and E. Nicoladis. 2007. Bilingual first language acquisition. In *Blackwell Handbook of Language Development*, ed. E. Hoff and M. Shatz, 324–342. Malden: Blackwell.
Grondin, J. 1995. *Sources of Hermeneutics*. Albany: State University of New York Press.
———. 1997. Die späte Entdeckung Schellings in der Hermeneutik. In *Zeit und Freiheit: Schelling – Schopenhauer – Kierkegaard – Heidegger; Akten der Fachtagung der Internationalen Schelling-Gesellschaft, Budapest, 24. bis 27. April 1997*, ed. I.M. Fehér and W.G. Jacobs, 65–72. Budapest: Éthos Könyvek.
———. 2000. Play, festival, and ritual in Gadamer: On the theme of the immemorial in his later works. In *Language and Linguisticality in Gadamer's Hermeneutics*, ed. L.K. Schmidt, 51–58. Lanham: Lexington Books.
Habermas, J. 1988. *On the Logic of the Social Sciences*. Trans. S. Weber Nicholsen, and J. A. Stark. Cambridge: MIT Press.
Hamacher, W. 1998. *Entferntes Verstehen: Studien zu Philosophie und Literatur von Kant bis Celan*. Frankfurt am Main: Suhrkamp.
Hammermeister, K. 2000. Heimat' in Heidegger and Gadamer. *Philosophy and Literature* 24 (2): 312–326.
Heidegger, M. 1976. Brief über den Humanismus. In *Martin Heidegger. Gesamtausgabe*, Wegmarken, ed. Friedrich-Wilhelm von Herrmann, vol. 9, 313–364. Frankfurt am Main: Vittorio Klostermann.
———. 1981. Erläuterungen zu Hölderins Dichtung. In *Martin Heidegger. Gesamtausgabe*, ed. Friedrich-Wilhelm von Herrmann, vol. 4. Frankfurt am Main: Vittorio Klostermann. English edition: Heidegger, M. 2000. *Elucidations of Hölderlin's Poetry*. Trans. K. Hoeller. New York: Humanity Books.
———. 1983. Gesamtausgabe. In *Aus der Erfahrung des Denkens 1910–1976*, ed. Hermann Heidegger, vol. 13. Frankfurt am Main: Vittorio Klostermann.
———. 1993. *Basic Writings*, Rev. Edn., ed. David Farrell Krell. San Francisco: Harper.
———. (2010). *Being and Time*. Trans. J. Stambaugh, Rev. D. J. Schmidt. Albany: State University of New York.
Meisel, J.M. 2006. The bilingual child. In *The Handbook of Bilingualism*, ed. T.K. Bhatia and W.C. Ritchie, 91–113. Malden: Blackwell.
Merleau-Ponty, M. 1968. *The Visible and the Invisible*. Trans. A. Lingis. Evanston: Northwestern University Press.
Michelfelder, D.P., and R.E. Palmer, eds. 1989. *Dialogue and deconstruction: The Gadamer-Derrida Encounter*. Albany: State University of New York.
Ricœur, P. 2008. *From Text to Action: Essays in Hermeneutics II*. Trans. K. Blamey and J. B. Thompson. London: Continuum.
Risser, J. 2012. *The Life of Understanding: A Contemporary Hermeneutics*. Bloomington: Indiana University Press.
Rotaru, I. 2010. Die ethische Priorität des Außerordentlichen: Interview mit Bernhard Waldenfels. *Studia Phaenomenologica* 10: 253–269.
Schelling, F.W.J. 1977. *Philosophie der Offenbarung: 1841/42*. Frankfurt am Main: Suhrkamp.
———. 1996. *Philosophie der Mythologie: In drei Vorlesungsnachschriften*, ed., K. Vieweg and C. Danz. München: Wilhelm Fink.
———. 2000. *The Ages of the World*. Trans. J.M. Wirth. Albany: State University of New York. German edition: Schelling, F.W.J. 1946. *Die Weltalter: Fragmente (In den Urfassungen von 1811 und 1813)*, ed. M. Schröter. München: Beck.
———. 2007. *Historical-Critical Introduction to the Philosophy of Mythology*. Trans. M. Richey, and M. Zisselsberger. Albany: State University of New York.
Schmidt, L.K. 1995. The ontological valence of the word: Introducing "On the Truth of the Word". In *The Specter of Relativism: Truth, Dialogue, and Phronesis in Philosophical Hermeneutics*, ed. L.K. Schmidt, 131–134. Evanston. Northwestern University Press.
Sloterdijk, P. 1988. *Zur Welt kommen – Zur Sprache kommen: Frankfurter Vorlesungen*. Frankfurt am Main: Suhrkamp.

———. 1993. *Weltfremdheit*. Frankfurt am Main: Suhrkamp.
———. 2013. *Ausgewählte Übertreibungen: Gespräche und Interviews (1993–2012)*, ed. Bernhard Klein. Frankfurt am Main: Suhrkamp.
von Humboldt, W. 1998. *On Language: The Diversity of Human Language-Structure and its Influence on the Mental Development of Mankind*. Trans. P. Heath. Cambridge: Cambridge University Press.
Waldenfels, B. 1987. *Phänomenologie in Frankreich*. Frankfurt am Main: Suhrkamp.
———. 1994. *Antwortregister*. Suhrkamp: Frankfurt am Main.
———. 1997. *Topographie des Fremden: Studien zur Phänomenologie des Fremden 1*. Frankfurt am Main: Suhrkamp.
———. 1999. *Vielstimmigkeit der Rede: Studien zur Phänomenologie des Fremden 4*. Frankfurt am Main: Suhrkamp.
———. 2010. Fremderfahrung, Fremdbilder und Fremdorte. Phänomenologische Perspektiven der Interkulturalität. In *Interkultur – Jugendkultur: Bildung neu verstehen*, ed. A. Hirsch and R. Kurt, 21–35. Wiesbaden: VS Verlag für Sozialwissenschaften.

Virtual Places as Real Places: A Distinction of Virtual Places from Possible and Fictional Worlds

Tobias Holischka

Abstract The development and spreading of personal computers has changed our lives in many ways. One of them is the notable fact that we are confronted with *virtual reality* at an increasing rate. At first glance, one might think about video games and similar forms of entertainment, but meanwhile many other technology-related aspects are virtual as well, like social networks, email accounts, and even bank accounts. Some are part of our lifeworld, others are perceived to be 'unreal' – fictional, illusory, deceptive. A specification of the ontological status of virtual environments seems necessary. Therefore, we need to state more precisely what we're actually speaking of by virtual reality. An interesting approach arises from thinking the virtual in terms of *place*: *Where* do video games take place? *Where* exactly is money in a bank account? *Where* are private email conversations stored? Virtual places are a suitable concept to explore computer-generated virtual reality, as it is independent from any specific content and hence allows a fundamental analysis. On this basis, we can compare virtual places with one another, with other non-material places like fictional places, and correlate them with the places of our lifeworld. This paper investigates the ontological status of virtual places by distinguishing them from fictional places and merely (logically) possible worlds. Virtual places thereby prove to be limited by consistency and human imagination, but also offer new features like interactivity and direct referencing of other users. As a spatial foundation for actual actions, they are part of our lifeworld.

Keywords Place · Virtual · Reality · Possibility · Potentiality · Modal logic · Fiction · Video games · Philosophy

T. Holischka (✉)
Catholic University of Eichstätt-Ingolstadt, Eichstätt, Germany
e-mail: tobias.holischka@ku.de

T. Hünefeldt, A. Schlitte (eds.), *Situatedness and Place*, Contributions To Phenomenology 95, https://doi.org/10.1007/978-3-319-92937-8_10

1 Actuality and Potentiality

The theory of possible worlds is based on the distinction between actuality and potentiality. Parmenides argued about 500 B.C. that being 'is' and not-being 'is not', opposing Heraclitus' ontology of becoming. The dispute over being and becoming ended with Aristotle's objection that there can be something even if it is not. A cherry stone for example includes the potential to become a cherry tree. This potential is part of the cherry stone, as it cannot become something different. As an aspect of the growing process, this potential is real, although it is not actual. Furthermore, potentiality does not necessarily become actualized at all. That means that what we refer to as reality is obviously more than just things and facts, it also includes mere potentiality (cf. Hartmann 1966, pp. 3–13).

In modern times, when René Descartes' philosophy prevailed over Aristotle's teachings, possibility replaced potentiality. Actual being became just one mode of existence, while potential being is another. Immanuel Kant points that out in his famous example of 100 coins: The value of 100 imaginary (i.e. possible) coins is the same as of 100 actual coins, just their mode of existence is different. Hundred possible coins are not less worth, they are just not physically present (cf. Kant 2009, B 106).

According to Kant, the category of modality contains three modes: *Possibility* means that something is merely possible, e.g. that a car can have various possible colors. *Contingency* is actuality by chance, which means for example that a specific car has actually a certain color. *Necessity* is the negation of contingency, meaning that something is not just possible, but also necessary, e.g. that a car has at least some color at all.

This concept of possibility offers an interesting perspective on computer-generated virtual reality. Virtual environments are, like not-actualized possibilities, not physically present, yet both exist in a certain way. If what we call reality, understood as actuality, just refers to one mode of existence, then also virtual phenomena seem to be no less real (cf. Welsch 1998, p. 200). For further investigations on their ontological status, another step into the history of philosophy is useful.

1.1 Possible Worlds

Gottfried Wilhelm Leibniz developed an interesting thought, proposing that every possibility is realized in a distinct possible world. According to him, God fulfils the realization of the best of all possible worlds by choosing from all possible, but among each other incompatible worlds, which is the one we call 'our' world. (cf. Leibniz 1985) Saul A. Kripke combined the concept of possible worlds with modal logic to relate them to each other. He understands modal operators as quantifiers of assertions on possible worlds: The existential quantifier corresponds here to possibility, while the universal quantifier equals necessity. If a certain thing is possible, that means, according to Kripke, it is actual in at least one world. In other words, there is at least one world where it is actual. If this world is coincidentally 'our' world, we call it *real*. Impossibility indicates that something is not actual in any world, while necessity denotes actuality in all worlds.

What we learn from Kripke is that the creation of stipulated worlds takes place by conditional clauses: *If* the world basically stays the same, however just one single fact is changed, a new possible world is specified. No matter how many items are modified, every possible world remains connected to 'our' real world as its origin and pattern.

If reality contains more than just actual facts, as we have seen before, we can now state with Kripke that there is an enormous number of possible worlds, as each possibility constitutes a new one. However, it is important to emphasize that Kripke's possible worlds are *stipulated*, they cannot be discovered by powerful telescopes (cf. Kripke 1980). We are bound to our sole world and cannot physically reach out to any other.

1.2 The View Through the Telescope

When it comes to virtual environments, we find a surprisingly similar construction. These computer-generated illustrations do, like possible worlds, originate in 'our' world – technically, as well as regarding their content. Computer simulations for example depict scenarios that are not actually happening somewhere. However, a different way of understanding them is Kripke's telescope metaphor: virtual worlds can be described, from that perspective, as interactive renderings of potential worlds (cf. Holischka 2016, pp. 58ff.). While mere potentiality is just accessible by thought, by thinking of what *could* be, computer technology helps us to express it in a visual and interactive way. Of course, video games, which are also a certain kind of virtual worlds, involve a lot of fictional content, and they should absolutely be looked upon as games. Nevertheless, they present a possible world in a virtual manner, no matter up to which extent they are differing from 'our' world. Possibility comprises everything that is possible, which also includes almost all video game scenarios, as long as we perceive their story as a part of a game.

This raises the question about the limits of possibility and the distinction from impossibility. David Lewis extends Kripke's theory on possible worlds: His modal realism assumes that possible worlds are not just stipulated, but rather they do actually exist (cf. Lewis 1986, pp. 84ff.). His key issue is to examine if and why impossible worlds cannot exist. According to Kripke, he claims, essential contradictions are not a problem for possible worlds, as long as they are just stipulated. In that case, they simply contain paralogisms on the logical level. However, asserting that inconsistent possible worlds do actually exist, violates Aristotle's law of non-contradiction

(Metaphysics IV 3, 1005b). It not only says that a statement is true and false at the same time, but furthermore claims that something exists and does not exist at the same time. That way, Lewis turns logic into ontology and makes plausible that impossible worlds cannot exist.

Lewis' thought is quite controversial, but it reveals a new perspective on virtual worlds: As they are not just imaginary, but rather tangible, they cannot illustrate impossible worlds. This puts a limit on their extent. If we follow Lewis so far, virtual worlds can solely depict consistent possible worlds. Obviously, we can find many contradictions in the storylines of video games, but these are just inconsistent narratives and remain on a semantic level. Of importance is what I call *ontic* consistency. We can e.g. stipulate a possible world where all circles are square (Kripke), but we cannot argue that there is another actual world where the same things are circular and square at the same time (Lewis). For the same reasons, we cannot create a virtual world where circular things are square. This world simply will not work as a world. It might look like a visual trick or like a work of M. C. Escher, but we just cannot accept its ontic contradictions. Virtual worlds must follow the rules of logic to be recognized as worlds.

Admittedly, this logical approach on determining virtual worlds has its pitfalls. Lewis' thought is by far not unquestioned and there are other logical systems that do not apply here. On the other hand, ontic consistency remains a criterion for virtual worlds that puts them into a new perspective.

2 Between Possibility and Fictionality

Not only possible and virtual worlds originate in our lifeworld, also fictional worlds do. This concept is quite similar, as we have seen before: Authors of fictional worlds cannot explain all inner-world references, so they just focus on differences to 'our' world (cf. Ingarden 1979). London, for example, as we know it from Doyle's *Sherlock Holmes* novels, premises on the real city, but mentions differences, like Mr. Holmes' house on Baker Street (cf. Haller 1986, pp. 71 f. & 82). Like possible worlds, fictional worlds are not independent, but always relate to our lifeworld in some way. Thus, the question arises if fictional and possible worlds are congruent. As we have seen with Lewis, possible worlds are limited by consistency, which however does not apply to fictional worlds. They can be (and usually are) full of contradictions, because they keep their focus on a good story. That is not a problem, as long as we consider this to be narratives and do not claim that fictional worlds are real.

At a first sight, it seems that all possible worlds can be imagined and therefore be couched as fictional worlds. Moreover, as fictional worlds can easily be inconsistent, fiction seems greater than possibility from this point of view. However, there is a practical limit in the creation of fiction, which is human imagination: If we cannot conceive something, it cannot become a part of a fictional world. The conclusion is that not all possible worlds can be fictional, and not all fictional worlds are possible. The overlap of both is where virtual worlds are located.

3 Virtual Places

Virtual worlds are subjected to the restrictions of possible and fictional worlds, but they expand both spheres fundamentally by being an *accessible interactive place* (cf. Holischka 2016, chapter 3). Possible and fictional worlds are static to some extent. While possible worlds are just notions and therefore do not include any tangible internal development, fictional worlds are bound to the author's narratives. Readers (novels) or spectators (movies) cannot decide what part of the worlds they would like to investigate, but need to take what the author tells them. Regarding that point, virtual worlds offer a fundamental frame for users to explore the world on their own and make changes that affect the further development of the whole world. Especially video games are a suitable example. That in turn requires a higher level of determination. While fictional worlds can dedicate a lot of detail description to the readers' imagination, virtual worlds need to depict these far more when it comes to a relatively free users' exploration of the virtual environment.

Another important difference is the facilitation of *direct mutual referencing* in virtual worlds. Readers of the same novel can experience a fictional world simultaneously, but they cannot 'meet' and act together. Virtual worlds, especially multiplayer video games, submit this feature. Players can meet in a virtual environment, interact (cooperate, trade, fight, etc.) and explore the world together. The referencing is not just general regarding other players per se, but direct and actual. As to that, virtual worlds are in experience more similar to our lifeworld than possible and fictional worlds can ever be.

The underlying basis of these differences is the concept of *place* (cf. Schlitte et al. 2014; cf. Holischka 2016). In the history of philosophy, place is often subordinated to time, as Edward Casey (1996) points out. We tend to ask first *when* something happened, instead of being aware that everything has its place and happens *somewhere*, which is often more meaningful than mere chronological order. Place is also competing with the more popular concept of *space*, which is a scientific abstraction, though. While place encompasses the qualitative experience of situatedness, space focuses on a quantitative model of distinctive elements in a theoretical matrix that reduces the content of experience to measurable scales.

> The crucial point about the connection between place and experience is not, however, that place is properly something only encountered 'in' experience, but rather that place is integral to the very structure and possibility of experience. (Malpas 1999, p. 31)

Edward Casey points out the ontological importance of place:

> The point is that place, by virtue of its unencompassability by anything other than itself, is at once the limit and the condition of all that exists. [...] At the same time, place serves as the *condition* of all existing things. […] To be is to be bounded by place, limited by it. (Casey 2009, p. 15)

Furthermore, he determines the nature of place as a happening:

> Rather than being one definite sort of thing – for example, physical, spiritual, cultural, social – a given place takes on the qualities of its occupants, reflecting these qualities in its own constitution and description and expressing them in its occurrence as an event: places not only *are*, they *happen*. (Casey 1996, p. 27)

To understand computer-generated virtual phenomena in a profound way, without discounting them as illusions or deceptions, we need to look at *where* they happen. While their contents derive from other contexts like communication or games, their specific characteristics as virtual entities remain spatial. It matters if a conversation takes place in a public chat room or in private mailboxes. Virtual worlds are, like possible or fictional worlds, spatial foundations of references, meaning and narratives. What distinguishes them from the others are their features of interactivity and direct user referencing. Hence, they found a basis for various actual activities. The immateriality of virtual places does not impugn their spatial virtue.

4 Instances of Virtual Placement

The term virtual place comprises several quite different phenomena, including basic computer applications, social media, simulations and even complex video games with persistent virtual worlds. To determine all of them appropriately, a new terminology needs to be introduced (cf. Holischka 2016, p. 22).

Trans-placement denotes a placement of place-concepts into virtual space that are commonly known from the lifeworld. For example, what we know as *desktop*, the first thing appearing when a computer has started up, is a functional imitation referring to the actual, material desktop, on which the computer is placed. Just as the conventional desktop is a place for paperwork, the virtual desktop grants similar functions regarding virtual file processing. *Trans-placement* is characterized by a recourse on widespread place concepts to depict computer functionality in a symbolic way, although other technical solutions would have been possible.

In contrast, virtual *in-placement* means a placement of concepts originating from the overlap of fiction and possibility in a virtual environment. As we have seen, both fiction and possibility are ontologically rooted in reality, but *in-placement* does not imply the importation of established place concepts, as *trans-placement* does, but the creation of *new places*, deriving from fiction or possibility as contentual sources and profoundly differing from the lifeworld. In this sense, especially contemporary video games contain new places.

While *trans-placement* and *in-placement* describe a varying creation of virtual places, they correspond reversely and at the same time with a tangible *re-placement* in material reality. This reference arises necessarily from the material foundation of every virtualization in pieces of technical equipment, such as computers, servers, storage media and datacenters. In this respect, virtual places are placed twice, on the one hand in virtual environments, on the other in technical systems that generate and host them.

All three aspects of virtual place are related to each other, but not in a circular way. Every virtual placement has recourse to the lifeworld, either as an analogy in the case of *trans-placement*, or when it comes to the entrenchment of possibility and fiction regarding *in-placement*, and generates new tangible places at the same time as their technical foundation. However, this does not constitute a circle of generating, because not every virtual place corresponds with a tangible place. Their relation is rather proportional, provided that an increasing level of virtualization leads necessarily to a more powerful technical background.

4.1 Trans-Placement

The initial task for designers of virtual environments was to translate completely new computer functionality into interfaces that users can understand and handle intuitively. With the implementation and spreading of graphical interfaces, especially for personal computers, only one implicit paradigm prevailed: to comprehend the computer as a workplace. The syllable 'work' derives from the fact that, due to the lack of processing power, complex multimedia applications were simply not executable, accordingly the focus within the early years was on text production and programming. More important in our context is that computers were seen as virtual places *in* which specific actions were to perform. The way to go was to import the well-known concept of an ordinary workplace into the new virtual environment in a symbolic manner. Consistently, we can find many lifeworld references with specifically spatial connotations. After turning on the machine, we find ourselves at a (virtual) desktop with (virtual) windows and (virtual) files in (virtually) accessible (virtual) folders. This kind of design is not necessarily the way it is. It is still possible to operate a computer just with plain text in terminal mode. Yet we prefer the graphical, spatial mode. What was meant to facilitate and enhance orientation and understanding, led to a spatial perception that we have to this day.

Even distinct applications stick to this paradigm. Word processors for example attempt to imitate the usage of conventional typewriters, along with additional functionality. There is no need to use virtual pages, but the concept of page is familiar to us, it is what we expect. Along with the desktop, word processors provide a whole environment in the sense of a virtual place that we can enter if we want to take notes. This is not a question of immersion – the application does tempt us to enter its own world or anything. We are always well aware that we interact with a machine, but we use it to reach out to a place that grants useful and comfortable functionality.

Based on this experience, many other lifeworld places were transformed into applications that serve as virtual places, for example traditional board games. The chessboard, for instance, is not just a board with 64 black and white squares in a characteristic array. It can also be seen as the place where chess matches happen, or, in other words, the chess matches as events constitute the checkerboard as a place of happening. The virtualization of the game transfers this place into another environment, but leaves the phenomenon intact. Additional features that come along with that, like fancy animations or computer opponents, do not matter in this perspective, as the game and the place of its happening stay essentially the same. This makes the virtual chess board a paradigmatic instance of *trans-placement*.

4.2 Virtual In-Placement

Not all virtual places derive their contentual origin from the lifeworld. Computer technology allows the creation of fantastic virtual worlds that seem to break the limits of physics, especially when it comes to video games. What makes them special in contrast to *trans-placement* is the fact that we cannot name lifeworld equivalents for them. This description applies even to less complex environments, like social networks. Technically, they can be characterized as account-based websites that provide personal information and audiovisual substance, along with distinctive features like direct communication and 'sharing' of content. This scheme has no template in the material world, yet it can be perceived as a phenomenon of place: we *enter* the network and interact *in* it. We even have virtual presence in it. I would not go so far as to proclaim that virtual interactions in social networks constitute something like a second, virtual life. Nevertheless, they are virtual places that bring people together and let them share their thoughts and feelings.

Much more sensational are video games. My favorite example is *Minecraft*, a so-called first-person-3D-sandbox-adventure, which we will examine later in detail. Contemporary video games comprise complex worlds that differ in many ways from our lifeworld. For one thing in terms of content: virtual environments offer the chance to escape the known setting, to use fictional elements to create fantastical and exciting worlds, and to establish a game context to try out devious roles and actions; For another thing regarding game mechanics: players do not die if their avatars do so, single levels or whole worlds may be instanced and can be restarted, stages of games can be saved and reloaded, and so on. Most of these points refer to the fact that we deal with games here, which differ from reality per se, merely secondarily that they take place in virtual environments. To get to the core of the subject, we need to cut out all aspects of game theory, which leaves us at the virtual places in which these games happen. Unlike the *trans-placement* of conventional games, these places are in-placed in terms of new virtual places that originate in possibility and fiction, not in material reality.

4.3 Tangible Re-placement

Trans-placement and *in-placement* are phenomenal aspects of virtual places that result from a technical approach. Virtual environments begin with theoretical conception, programming, design and practical implementation. Once the whole project is built up, users start to experience virtual places. These places are not constructed directly, but they emerge from a technical background and depend on it. This background changes after the creation process from the developers' computers to servers that sustain virtual environments. Unlike personal computers, servers are housed in datacenters, which can be considered as a kind of *Non-Places* (cf. Augé 1994): noisy and austere multi-level functional buildings that are optimized for servers regarding temperature, energy, security and fire prevention. Only a few technicians work there to keep the systems running. These technical places serve no other purpose than to provide the material foundation of virtual environments. From that point of view, virtual places cause backlashes into tangible reality by occupying corporeal places for their upkeep. We even may conjecture that virtual places are quantitatively limited on principle by availability of space and resources in our lifeworld.

This conjunction arises from the nature of information, which can be traced back into ancient Greece, especially the concept of matter and form. In this sense, information can be described as a bringing of form into matter (*in-formatio*), which binds the term to both of them (cf. von Weizsäcker 1979, pp. 52 f., and Capurro 1987, pp. 17–49). In information age, we can find this entanglement represented in software and hardware – the first corresponds to form, the latter to matter. As virtual places meet their technical basis in software, they necessarily require a material substrate.

Forms of *re-placement* co-occur with all virtual places. Complex and persistent virtual worlds demand permanently running powerful servers that occupy places, but, even in smaller scales, every digital information requires a storage medium. *Re-placement* comes to pass when we look at a virtual desktop or word processor, even as to virtual file management. Files are typically stored in virtual volumes with a folder hierarchy. We can drag and drop them into other folders or applications, and also copy, rename and delete them. The volume works as a kind of virtual container, and some of them are even portable: Thumb drives can be described as virtual places that carry data through the material world from one computer system to another. Technically, we cannot determine exactly where a single file is stored to a volume, as processes like safety redundancies, checksums and varying characteristics of different file systems are of importance. Virtually, however, we still assume that files are stored in the folder hierarchy.

The new *cloud*-technology pursues to break this affiliation, trying to shift and hide the aspect of *re-placement* into uncertainty. We do not need to care about *where* files are stored, as long as they are present when we need them, on all of our devices. Technically, they claim their place very well, they are just saved on internet servers and synchronized online. The unsettling question remains where these servers are located and who has access to them. Not only has this aspect of *re-placement* led to worrying questions of power and sovereignty.

Another category of phenomena of place is also to be classified as *re-placement*, which is the material places that grant access to virtual places. While servers are typically not directly accessible to users, personal computers are even more. A desk with a computer display and input devices like mouse and keyboard can be described as place of transition between worlds. Causation works in both directions here: On one hand, these places are used to create virtual places, but on the other, they are arranged with the special purpose of entering virtual environments. Moreover, due to Wi-Fi technology, they spread within our lifeworld. Many public places like campuses or cafes offer wireless internet connection, which gives them the second meaning of places of transition to virtual places. This is another meaningful example of the virtual encroaching on lifeworld.

5 The Virtual World of *Minecraft*

Virtual places appear most impressive when it comes to video games. In the context of *in-placement,* they roll out huge virtual worlds with fantastic stories, fascinating adventures and complex game mechanics. The aspect of place becomes even more relevant in multiplayer games where players can meet, interact and actually reference each other. A perfect instance is *Minecraft*, a sandbox construction game with multiplayer features. It has different game modes, but sufficient in our context is survival mode. This chapter is intended to give an example of a virtual world that allows creating and experiencing virtual places in a very explicit manner.

The game's basic attribute is that its whole world is built with blocks in the size of a cubic meter, arranged in a grid array, which gives the game an arbitrary look. Unlike other contemporary video games in that genre, it does not attempt to look like reality – players are constantly reminded that they interact within an artificial world, yet immersion sets in.

In *Minecraft*, everything is made out of blocks, like pasture, lakes, trees, mountains, animals and even the player's avatar. We find different landscapes like grassland, deserts, jungles and arctic zones, and all are seamlessly connected. Each *Minecraft* world has a theoretical size of 3,6 billion square kilometers (earth has about 510 million), depending on the players' exploration activity, which leaves plenty of space for huge projects. The game has no specific goal to accomplish, therefore the players are free to do whatever they like, typically building houses, farms, tunnels, and a lot more. To do so, the player breaks blocks and places them on each other. Blocks represent different materials with distinct properties, depending on where they are harvested, e.g. tree trunks, stone, sand, water. These materials can be transformed to create tools, bricks, planks, glass and many more. Therefore, the basic activities are *mining* and *crafting* to produce whatever is necessary to complete a project. What sounds quite easy turns out to be really complex. Special matter like *Redstone* for example allows to construct circuits to control automated processes like railroad networks and even rudimentary computers. However, it takes a lot of time to get there.

5.1 *The First Nights*

When the game starts in adventure mode, the player is thrown into the block-world in the first-person-perspective of his avatar. Although the world presents itself in blocks, the meaning of its arrangement is evidently a scenario of (artificial) nature. The player gets no instructions what to do, so he starts exploring the world and realizes soon that he has the ability to interact with it, which basically means breaking blocks, storing them in the inventory, and placing them again into the world.

While he is playing around, darkness is falling, and with it, scary creatures arise and start attacking the player. A daring escape follows, the player tries to survive, stampeding through a dark world that makes him loose orientation soon. After endless minutes, this unsettling situation ends with daybreak, the persecutors burst into flames and disappear. The player realizes that this world is hostile and full of terrors, especially at night. Unburdened sandbox activities come to an end, a save place for the night is to be found. Leaving the game does not solve anything, as the stage of the game is saved automatically and reloaded upon restart.

From a beginner's point of view, the easiest way to avoid confrontation with scary creatures is to break blocks, dig a cave into the ground, seal it with replaced blocks and simply hide. The cave is dark, narrow and boring, which is why the player starts experimenting with his inventory to kill time. The inventory contains all blocks that have been broken so far, which is, at this stage of the game, usually dirt, sand and wood, arranged in stacks up to 64. A separate field of 2x2 squares is denoted as 'Crafting', which can be used to transform natural blocks into new blocks, e. g. wood into wood planks. The player has no guide about crafting and is therefore depending on trial and error. After some time, he learns that wood planks can be carved into sticks. Sticks again can be used to create wooden pickaxes. This tool is essential to break more solid blocks like stone. Digging deeper into earth, the player will find coal, which can be used to craft torches to create artificial light. Illumination is essential, both for orientation below ground and to keep nightly creatures at distance.

However, another difficulty arises soon: The avatar's state of health is constantly decreasing. No obvious reason is given for that, but in perspective of similar games, the lack of food could be the cause. In the course of world exploration, edible items like fruits or vegetable might already have accidentally been picked up and can be consumed in the inventory, which defuses the situation for the moment. Yet, a reliable source of food is to be found. Experimenting with crafted tools and natural elements of the virtual world offers ways to breed cattle, cultivate fields and prepare food. To survive does not exhaust itself in avoiding danger, but also in actively taking care of the avatar's diet. At this point, the player has a first save haven and rudimentary tools for further exploration of the world of *Minecraft*.

5.2 Dwelling in a Virtual World

The process of crafting is too complex for a detailed description in this context, but it can be summarized as a dynamic system of a technological process to transfer natural blocks into artificial blocks and specialized tools that again grant access to scarcer blocks and new comprehensive functionalities like blacksmithing, brewing and even magical enchanting. As the game goes along, the initial task to survive in a hostile world is fading into the background. In heavy armor and equipped with powerful weapons, it is no more a big deal for the avatar to defeat vicious nightly creatures, and as illumination keeps them off, it is unnecessary to fight them anyway. Food supply is covered with farms and plantations. Tamed horses allow fast travel and vast mines grant access to rare materials. Livelihood is more than secured, and that is where the object of the game shifts.

The player realizes that his previous steps were something like an implicit tutorial to the game, to get familiar with its procedures and functionalities. He arrives in a true sandbox game that does not provide tasks or quests anymore, but gives the freedom to do whatever is intended. Typically, a phase of home improvement breaks loose: Why live in a functional brutalist bunker? In the mining process, a lot of cobblestone blocks accumulate and are usually used to build a first shelter. But they can be smelted into stone blocks that simply look better. The paradigm of survival changes to dwelling, to virtually make oneself at home in the virtual world of *Minecraft*. Torches are replaced by lamps, stairs and doors are installed, and even pictures and bedside rugs come to use to beautify rooms. All of this has no relevant purpose in the game, but it allows the player to create a place that feels like a virtual home.

The bedroom carries great weight in this constellation. The functions of the bed are to 'sleep', which effectively means to skip nighttime, and to set a dedicated respawn point in case of the avatar's virtual death. In both cases, a perfectly save bedroom is absolute necessary, as sleeping will not be possible if hostile creatures are close, and also to prevent immediate death after respawn. To ensure this, the bedroom is usually in the center of the most important building, well illuminated and secured with the most resistant materials. Typically, valuable tools and rare blocks are kept close, and most effort in decoration is elaborated here. This makes the bedroom one of the most relevant places in the game, the center of the virtual home. All activities, like construction projects and extended explorations, originate and come to an end at this place. When leaving the game, players prefer the bedroom to do so, as it guarantees a save reentrance. The bedroom in *Minecraft* is not only a paragon of a virtual place, it is also the place of entry into its virtual world.

6 Conclusion

As we have seen, virtual places can be qualified as phenomena of place, even if they are only accessible with the help of computer technology. Place appears to be an interesting concept to understand virtual entities and to connect them ontologically to our lifeworld. It is important to emphasize that virtual worlds strictly depend on what we call reality, especially regarding *trans-placement, in-placement* and *re-placement*. As Jeff Malpas points out (2009), the virtual does not only physically and causally depend on what we call reality, but also as regards content. Furthermore, the limitations of the users' / players' embodiment tie them to their lifeworld, as well as their language and everything that is of meaning to them. "[I]t may be taken to suggest that the virtual is *different* from the everyday, whereas strictly speaking, the virtual is merely another part or aspect of the everyday world [...] There is thus only the one world, and the virtual is a part of it." (Malpas 2009, p. 135) Given that, we can conceive virtual places as an extension of the lifeworld into the virtual. Hereby, the phenomenon of place serves as a structure that combines both spheres, as it exonerates virtual places from the reproach of being illusionary or deceptive. We should consider them a part of our lifeworld.

References

Augé, M. 1994. *Orte und Nicht-Orte. Vorüberlegungen zu einer Ethnologie der Einsamkeit.* Frankfurt am Main: S. Fischer.

Capurro, R. 1987. *Information. Ein Beitrag zur etymologischen und ideengeschichtlichen Begründung des Informationsbegriffs.* Munich/New York/London/Paris: K.G. Saur.

Casey, E.S. 1996. How to get from space to place in a fairly short stretch of time: Phenomenological prolegomena. In *Senses of Place*, ed. S. Feld and K.H. Basso, 13–52. Santa Fe: School of American Research Press.

———. 2009. *Getting Back into Place. Toward a Renewed Understanding of the PlaceWorld.* 2nd ed. Bloomington/Indianapolis: Indiana University Press.

Haller, R. 1986. Wirkliche und fiktive Gegenstände. In *Facta und Ficta. Studien zu ästhetischen Grundfragen*, ed. R. Haller, 57–93. Stuttgart: Reclam.

Hartmann, N. 1966. *Möglichkeit und Wirklichkeit.* Berlin: De Gruyter.

Holischka, T. 2016. *CyberPlaces. Philosophische Annäherungen an den virtuellen Ort.* Bielefeld: Transcript.

Ingarden, R. 1979. Konkretisation und Rekonstruktion. In *Rezeptionsästhetik. Theorie und Praxis*, ed. R. Warning, 42–70. Munich: Fink.

Kant, I. 2009. *Critique of Pure Reason.* Cambridge: Cambridge University Press.

Kripke, S.A. 1980. *Naming and Necessity.* Harvard: Harvard University Press.

Leibniz, G.W. 1985. *Theodicy: Essays on the Goodness of God, the Freedom of Man, and the Origin of Evil.* Trans. E.M. Hubbard. Lasalle: Open Court.

Lewis, D.K. 1986. *Counterfactuals.* Oxford: Blackwell.

Malpas, J.E. 1999. *Place and Experience. A Philosophical Topography.* Cambridge: Cambridge University Press.

———. 2009. On the non-autonomy of the virtual. *Convergence* 15: 135–139.

Schlitte, A., T. Hünefeldt, D. Romić, and J. van Loon. 2014. *Philosophie des Ortes. Reflexionen zum Spatial Turn in den Sozial- und Kulturwissenschaften.* Bielefeld: Transcript.

von Weizsäcker, C.V. 1979. Sprache als information. In *Die Einheit der Natur. Studien von Carl Friedrich von Weizsäcker*, ed. C.V. von Weizsäcker, 39–60. Munich: Hanser.

Welsch, W. 1998. Eine Doppelfigur der Gegenwart. Virtualisierung und Revalidierung. In *Medien-Welten. Wirklichkeiten*, ed. W. Welsch and G. Vattimo, 229–248. Munich: Fink.

Situated Anxiety: A Phenomenology of Agoraphobia

Dylan Trigg

Abstract Anxiety is sometimes thought of as either a state of mind, lacking a thick spatial depth, or otherwise conceived as something that individuals undergo alone. Such presuppositions are evident both conceptually and clinically. In this paper, I present a contrasting account of anxiety as being a situated affect. I develop this claim by pursuing a phenomenological analysis of agoraphobia. Far from a disembodied, displaced, and solitary state of mind, agoraphobic is revealed as being thickly mediated by bodily, spatial, and intersubjective dimensions.

Keywords Anxiety · Place · Agoraphobia · Intersubjectivity · Bodies · Spatiality

1 Introduction

In the history of philosophy, no affective state is more venerated than anxiety. From Seneca and Epictetus to Kierkegaard to Heidegger (to say nothing of psychoanalysis), anxiety has assumed a place in conceptual thought as a state that is both disclosive of fundamental truths and transformative of human agency in equal measure. To think here of the three modern icons of anxiety; namely, Kierkegaard, Sartre, and Heidegger. Each of these thinkers invokes anxiety in one way or another as a mood that brings about fundamental change in human existence. Whether it is Kierkegaard's compelling and forceful account of anxiety as a dizzying force that can ultimately educate us, or Sartre's incisive analysis of anxiety as the anguish of our freedom, or Heidegger's account of anxiety as the means through which an authentic rapport with death is made possible, anxiety is a pivot around which many twentieth century philosophies revolves.

Yet for all its centrality, anxiety remains oddly abstract in its conceptualization. In both philosophical research and clinical studies, anxiety is often presented as either a state of mind, a placeless experience, or otherwise construed as something that individuals undergo alone. Such a characterization leaves to one side the richly

D. Trigg (✉)
University of Vienna, Vienna, Austria
e-mail: dylan.trigg@univie.ac.at

T. Hünefeldt, A. Schlitte (eds.), *Situatedness and Place*, Contributions To Phenomenology 95, https://doi.org/10.1007/978-3-319-92937-8_11

situated quality of anxiety. In this chapter, we wish to counter this prevailing account of anxiety by demonstrating how anxiety is situated in a multifaceted sense. We take the term "situated" to refer to how spatiality, embodiment, and intersubjectivity are neither contingent nor incidental aspects to the experience and understanding of anxiety, but are constitutive of the mood in the first instance. Situatedness in this sense underscores the multidimensional experience of anxiety long before it is conceptualized in abstraction. In this respect, while our focus is on anxiety, the insights developed from this study are illustrative of how affective states more generally are situated in a relational and affective sense.

Our discussion begins by charting the manner in which anxiety is presented in philosophical and clinical literature. After exploring some of the salient ways in which anxiety is understood, we then proceed to outline in a positive sense the situated structure of anxiety. Our case studies consist of looking at the different modalities of situatedness in agoraphobic anxiety. Agoraphobia is an especially rich condition, which concerns a multiplicity of spatial anxieties irreducible to a fear of public space, and that alone. Moreover, agoraphobia presents us with an especially clear (and often striking) sense of how both anxious and non-anxious existence is structured by bodily, spatial, and intersubjective aspects. Agoraphobic anxiety also reveals how a disruption in bodily existence can also lead to disruption in a sense of self more broadly. This is clear in at least two ways. In the first case, the anxiety specific to agoraphobia often involves a disturbance in bodily motricity, such that the sensation of anxiety, including the inability to move or the sudden urge to move, is felt as if it comes from nowhere. In the second case, the body of the agoraphobic person is often presented as a distinct *thing* in the world rather than a center of selfhood or a body-as-subject (cf. Trigg and Gallagher 2016).

This richness inherent in agoraphobia lends itself especially well to demonstrating the situated quality of anxiety. In what follows, we will look at three such situated dimensions: bodily, spatial, and intersubjective. As we will see, the situated dimensions are not isolated from one another, but instead co-constitutive in a mutually edifying manner. In a chapter of this sort, it is impossible to do justice to the complexity of a condition such as agoraphobia, and a lengthier analysis of agoraphobia can be fruitfully pursued elsewhere (Trigg 2016). Our aim, rather, is more modest: to argue for the situated quality of anxiety by presenting an overview of agoraphobia.

2 Heidegger on Anxiety

We begin by considering some of the ways in which anxiety is conceived philosophically and clinically. Given the complexity of the topic, it is beneficial to narrow our analysis to a specific formulation of anxiety. To this end, we will turn toward Heidegger. Heidegger's account of anxiety is venerated as an incisive analysis of how the disruption of our habitual, taken-for-granted experience of everyday existence leads to a renewal of ontology and thus to the possibility of authentic being

(Heidegger 2008). His elevation of anxiety is unrivaled in terms of both its epistemic and existential status. Heidegger's analysis takes as its point of departure an account of human existence, whereby the world is tacitly given as a whole. Nothing is questioned because the world is evidently just *there*. Heidegger thinks that this conflation of presence and self-evidence neglects the world as a foundational – ontological – ground of our existence. Such a world is not one that can be grasped in an objective sense; rather, it is the source of the very action of grasping itself. Certain moods, however, bring to light aspects of existence that are ordinarily overlooked. One such mood is anxiety. Anxiety pierces through the habitual taken-for-granted relation we have with the world, confronting us in no uncertain terms with the "nothingness" that is constitutive of existence (Heidegger 1977, p. 103). Anxiety is thus, in a minimal sense, a mood that serves a methodological role in Heidegger's phenomenology. But anxiety is, as Heidegger also suggests forcefully, a mood that best captures the uncanniness of our existence, even when we are not explicitly aware that such uncanniness accompanies us.

The reason that Heidegger elevates anxiety in such a way is because it is marked out as an "original mood," original because it confronts us with the fact that "they are beings – and not nothing" (Heidegger 1977, p. 105). How does anxiety do this? Heidegger suggests that during anxiety, the meaning we assign to things in their everyday context slips away, including that of our own selves. In this rupture of things, the contingent dimension of meaning comes to the foreground. The result is that we feel "ill-at-home" in the world, as the world reveals itself in its strangeness (Heidegger 1977, p. 111). Moreover, alongside this ontological dimension, anxiety carries with it a transformative structure. What this means is that anxiety – in some capacity – instigates a fundamental insight into the nature and structure of the subject. According to Heidegger, anxiety is privileged not only for giving us access to things in their totality, but also for having the potential to redefine a more authentic relation to those things, as he has it anxiety is not one mood among many, but is instead a "fundamental kind of attunement belonging to the essential constitution of Da-sein" (Heidegger 1977, p. 177). By entering into a relationship with anxiety, other possibilities of being-in-the-world are opened up, and without the confrontation with anxiety, Dasein's eventual approbation of its own existence remains impossible.

We have considered Heidegger's account of anxiety in order to get a sense in the broadest fashion of some salient ways in which the mood is treated. Accordingly, we might divide these tendencies into three categories, each of which has implications for a situated account of emotions: placeless, disembodied, and individualistic.

Placelessness Heidegger often describes anxiety as a state in which we "hover" and "more precisely, anxiety leaves us hanging because it induces the slipping aware of beings as a whole" (Heidegger 1977, p. 103). Anxiety's affective force is due in part to its nebulous character; it is a presence that is at once an absence, and which resists being placed, as he says in characteristic terms "it is already 'there,' and yet nowhere" (Heidegger 2008, p. 231).

Disembodied For Heidegger, anxiety is a state that strips us of our bodily expressivity. Nowhere is Heidegger's notoriously missing body more conspicuously absent than in his account of anxiety. His is a body that is silenced, suppressed, and, moreover, defined solely as a "state-of-mind" (Heidegger 2008, p. 232). Being anxious means being in an anxious state-of-mind about "Being-in-the-world," such that the world slips away from us. Anxiety, he remarks, "robs us of speech" (Heidegger 1977, p. 103). Anxiety's placelessness is thus tied up with its disembodied structure. If human existence remains *in* the world during anxiety, then it nevertheless is existentially uprooted from the world, defamiliarizing that which is the most familiar of things, not least the body.

Individualistic Heidegger's affirmation of anxiety carries with it an implicit qualitative value. That anxiety strips us of our taken-for-granted complicity with the world means that human existence is forced to redefine its relation to the world, including relations with others. As he sees it, our inauthentic and largely non-anxious rapport with others consists of an absorption in the generalized and anonymous voice of the many. It is not that *I* have such and such a belief about the world, but that *they* do, and thus I form a compact with this anonymous one. The fulfillment of Heidegger's project, then, consists of employing anxiety as the means for human existence to undergo a transition from absorption in the anonymous masses to a state of individuation, marked by an anxious affirmation of death as *my own*. To this end, the value of anxiety of Heidegger is framed at all times by its power to transform Dasein from one mode of being to another. While it is true, that this analysis is remarkably insightful in terms of giving form to anxiety's uncanny affect, Heidegger nevertheless remains committed to an account of anxiety that is conceptual in focus, and thus neglects the situated structure of the mood.

3 Clinical Literature

We present this brief sketch of Heidegger in order to delineate some trends not only in his thinking but in the characterization of anxiety more generally. His analysis of anxiety is both incisive and subtle. Through his careful descriptions, Heidegger reveals how anxiety is far from a transient or contingent psychological state, but instead a mood that discloses fundamental aspects of existence. Moreover, his evocative descriptions of anxiety as an uncanny state shed insight on why anxiety is such a pervasive and powerful force, revealing, as it does, the precarious ground upon which meaning is instituted. It would require another treatment to delineate the flaws in Heidegger's account of anxiety, and for the present purpose we wish to only underscore the non-situated dimension of anxiety. Yet the framing of anxiety as placeless, disembodied, and individualistic is not limited to Heidegger. Rather, in the clinical literature on anxiety, albeit in a language radically different to that of Heidegger's, these same trends appear in another guise. The body, for example, while not wholly absent in this research, is nevertheless presented as a set of discrete functions, which purportedly give us an objective picture of anxiety. Let us consider here an account offered by leading anxiety specialist, David Barlow:

> Activation of the cardiovascular system is one of the major components [of anxiety]. Typically, peripheral blood vessels constrict, thereby raising arterial pressure and decreasing blood flow to the extremities. Excess blood is redirected to the skeletal muscles that can be used to defend oneself in a struggle. Blood pooled in the torso is more available to vital organs that may be needed in an emergency. Often people seem "white with fear"; that is, they blanch with fear as a result of decreased blood flow to the skin. Trembling with fear may be the result of shivering and perhaps piloerection, in which body hairs stand erect to conserve heat during periods of vasoconstriction" (Barlow 2002, p. 3).

Barlow goes on to provide an explanatory account of the various biological functions of the body during anxiety, ranging from tight breathing to the dilation of pupils to the occasional urge to vomit. His explanatory motivation is to underscore the "shadow of intelligence" and rationality of the organism in its capacity to adapt and respond to given dangers (Barlow 2002, p. 4). Anxiety serves to manage dangerous situations through activating neurobiological components of the body, which in turn raise levels of vigilance and enables subjects to anticipate dangers in the future through learning from past experience. If Barlow's rich and exhaustive research on anxiety is not reducible to framing the body solely in biological terms, then the treatment of the body as a unity of different process is nevertheless accented at the expense on a focus on the lived and relational dimensions of anxiety. Indeed, Barlow's focus on the neurobiology of anxiety is in keeping with recent trends in anxiety studies in attending to genetic, psychophysiological, and neurobiological data (Barlow 2002, p. 43).

The counterpart to this emphasis on the biological dimension is a focus on the cognitive structure of anxiety, evident as much in theoretical accounts as it is in treatment. A condition such as agoraphobia would, for example, seem to lend itself especially well to being treated by cognitive oriented behavioural therapy (CBT). One reason for the relevance of CBT is that symptoms of agoraphobia present themselves as discrete events in what is often otherwise a functional existence. Thus, with use of CBT, patients are "educated" about the physiological processes that give rise to an acute sense of anxiety. Once the subject "accepts" that their anxiety is a misinterpretation of perceived danger, "the secretion of adrenaline" is diminished thanks to a "cognitive restructuring" (Aslam 2012). Such an educational approach is often implemented alongside exposure therapy, where the patient is encouraged to desensitize themselves to places and situations that are liable to invoke and provoke anxiety (Edelman and Chambless 1993). Patients are then asked to repeat the procedure by way of "homework" in order to facilitate and expedite the desensitization process, until the patient is entirely acclimatized to the fact that the places originally thought of as terrifying are, in reality, devoid of danger. As a result, the patient is able to inhabit the world without the sense of impending collapse previously associated with venturing outside the home.

Such an understanding and treatment of anxiety is problematic in at least three respects. First, so long as the body is treated in mechanistic terms as an instrument to be reconfigured and retrained, then no attention is given to the situated and lived dimension of anxiety – not least, to how anxiety is structured in respect of our relation with other people and to the constitution of space through the lived body. This leads to the second problem. According to the clinical and cognitively orientated reports, the spatial dimension is at best a backdrop to anxiety and at worst irrelevant. Thus, from the perspective of a cognitive-behavioral model, spatiality is thought of as being a largely neutral canvas, an already formed container, against which the agoraphobic person needs to "restructure" their way of thinking (Martin and Dahlen 2005). Finally, throughout this analysis, the emphasis remains in large on individual experiences of anxiety. Such a focus is not peculiar to the cognitive-behavioral model, and indeed is evident in the philosophical and theoretical lineage on anxiety from Kierkegaard to present day memoirs on the topic. Take Rollo May's definition of anxiety an exemplar of this tendency:

> [Anxiety is] the apprehension cued off by a threat to some value that the individual holds essential to his existence as a personality. The threat may be to physical life (a threat of death), or to psychological existence (the loss of freedom, meaninglessness). Or the threat may be to some other value which one identifies with one's existence (patriotism, the love of another person, "success," etc.). (May 1996, p. 180)

This atomistic account of anxiety as an experience concerning the inner life of an individual certainly has value. But such a focus needs to factor alongside a more contextual analysis of anxiety. Anxiety does not simply concern the inner experience of one's individual existence; rather, it is framed at all times by a bodily, spatial, and intersubjective structure. In the remainder of the paper, we counter such an approach by turning toward a phenomenological analysis of agoraphobia in order to demonstrate the situated quality of anxiety.

4 Toward a Situated Account of Agoraphobia

Agoraphobia is commonly thought of as a spatial anxiety, especially a fear concerning public places (Knapp 1988, Vidler 2000). The term "agoraphobia" was coined by the nineteenth century German psychiatrist Carl Westphal. Deriving from the Greek word "agora," the phobia (from *phobos* meaning fear or dread) is in fact broader than that of the market place, and concerns as much the people gathered in that space together with the space itself as it does an apprehension of anxiety. Westphal himself observed symptoms of dizziness in several of his patients, yet was not convinced that these symptoms could be explained in physical terms alone. The feeling of anxiety in his patients, he suggests, "were more in the head than in the heart" (Knapp 1988, p. 60). Anxiety's focus, to put it in tautological terms, is nothing less than anxiety.

At the heart of this anxiety is a concern with being trapped in places where an escape would be difficult (Chambless and Goldstein 1982). Such places need not only refer to actual locations such as supermarkets, shopping malls, and motorways; it can also refer to situations where escape would prove difficult, such as being in the middle of a conversation with another person. These situations prove anxiety-inducing for agoraphobic people given that each situation has a tendency to produce distressing and unfamiliar sensations within the lived body (Trigg 2013a, 2016). The emergence of agoraphobic anxiety tends to follow a fairly uniform pattern. A person susceptible to anxiety is stricken by panic in a public place. Thereafter, they avoid that place and foster a more dependent relation to their home. In time, the avoidance patterns expands until the area of comfort available to them recedes to an increasingly more diminished point in space, often limiting their spatial existence to the immediacy of the home. If they are able to function in the world, then it is only in a highly ritualized and contorted way, avoiding places that spur feelings of unfamiliarity and loss of control while adhering to a predictable pathway that affords them the illusion of being in control. Physical sensations are consistent with those of a panic disorder: shortness of breath, trembling legs, palpitations of the heart, and in more intense cases, depersonalization and derealization (Chambless and Goldstein 1982, p. 2). More often than not, the symptoms of agoraphobia are partly alleviated when the subject is accompanied by the presence of a trusted companion or otherwise carries with him or her a familiar and supportive "prop," such as a pair of sunglasses, an umbrella, or any other totem of familiarity.

Each of the case studies in Westphal's original description testifies to these characteristics. We read of patients who "cannot visit the zoo in Charlottenburg, because there are no houses," who feel compelled to return home after reaching the "city limits," who think they are otherwise healthy aside from their agoraphobia, and indeed who are otherwise physically healthy (Knapp 1988). The behavior of the agoraphobic patient is marked by a cautious awareness of situations likely to produce anxiety, and is insistent on adhering to a familiar set of habits that he or she knows will lessen the risk of provoking such a response. Strikingly, the onset of agoraphobia is often met with a confusion on behalf of the patient, and in both Westphal's studies alongside contemporary reports, the patient cannot account for *why* he or she is anxious: "He is absolutely unable," so Westphal writes, "to offer a specific reason for his feeling of anxiety" (Knapp 1988, p. 66). With this overview, we aim to give an indication of how agoraphobic anxiety is fundamentally situated, insofar as it instead of being conceivable as a set of local symptoms, the mood is instead embedded in a nexus of subjective, spatial, bodily, and intersubjective structures. In what follows, we will unpack these situated dimensions by turning to a case study.

5 Anxiety as Bodily

The case concerns that of Allen Shawn, an American composer who has written lucidly on his own agoraphobia (Shawn 2007). Allen is multi-phobic, meaning that phobic anxiety affects his life in a complex and multifaceted sense. He is afraid of heights, of being on water, avoids bridges, is both agoraphobic and claustrophobic at once (as is often the case), avoids all forms of public transport, and loathes being in buildings without windows (Shawn 2007, p. xviii). If he is able to travel at all, then it is only "within a circumscribed world and without spontaneity" (Shawn

2007, p. xix). Many attempts are needed in order to "rehearse" departing and arriving at a place. Our aim here in turning to Shawn is to argue for the situated quality of anxiety. To achieve this aim, it is worth plotting a thematic counterpart to that of Heidegger, outlining the fundamentally bodily, spatial, and social aspects inherent in agoraphobia. We begin with the body.

To understand the specificity of the body as it plays a constitutive role in agoraphobic anxiety, consider the operation of the body in non-pathological experience. For the most part, we experience ourselves as unified agents. When we move from one corner of a room to another, then we do so both as the agents of our own movement and without having to question those movements. In a word, our *bodily motricity* remains at all times intact. We understand bodily motricity as the body's power to project into the world as a movement of spontaneity and possibility (Merleau-Ponty 2012). What this means is that when we move through the world, then we do so not as if the world were a series of dissected chunks that we have to manage piece by piece. The world, by contrast, presents itself to us as one unified whole, such that there is an intimate liaison between body and world. The implication is that our body operates according to a certain logic, which, whilst not always available to us in reflection, nevertheless serves to underscore a temporal and spatial unity operational "beneath intelligence and perception" (Merleau-Ponty 2012, p. 137).

The case of agoraphobia is different. To get an immediate sense of just how different agoraphobia is, let us consider Allen Shawn's trouble walking down a road "bounded on both sides by large open fields":

> I don't have too much trouble walking many miles at a time in busy New York City, but when I got halfway down this empty road, I would freeze in place and balk at continuing, exactly like a dog who freezes at the door to the veterinarian's office or a horse who refuses to walk over a rotten bridge. I couldn't be convinced that I could continue to walk despite whatever symptoms I felt and that if I did so, I would in fact get to the end of the road and still be the person I was four-tenths of a mile back [...] I oriented myself by seeing how far I had walked from one small tree to another, but at a certain point, less than halfway down the road, I would stop and simply wouldn't budge [...] I was convinced that when I reached the midpoint of the road, my legs would not move at all that I would be trapped in place there ... I should mention that in order to undertake this tiny journey, I had come equipped with all my 'safety items' [...] My friend was astonished. She tried to coax me, offered a kiss as a reward, promised not to leave me stranded (Shawn 2007, pp. 117-118).

Note how Shawn's troubles do not consist in a fault in his physiology. When in New York City, one thing is clear: he is able to walk. This ability is thanks in part to the city's familiarity and lack of open spaces. To understand the development and meaning of his anxiety, it is not enough to assume a third-person perspective on the operations and processes of Shawn's body considered as a set of mechanisms. Moreover, his anxiety cannot be understood as a set of numerical levels and measurable variables. Rather, we have to understand how both the body and spatiality co-shape each other, and how the situated body is central to the manifestation and onset of anxiety. The "measure" of Shawn's anxiety is not quantifiable. We must, instead, think alongside the experience of anxiety in a relational manner. Far from a disembodied state, as Heidegger and cognitively oriented research would suggest, anxiety can only be understood in and through the lived dimension of the body.

In Shawn's illustration, we witness a circular dynamic that is central to the development of anxiety. There are two points to mention. The first point concerns the expressivity of the body. When we are engaged in a meaningful act, then for the most part, we feel that our bodies are expressive of subjective experience. We express ourselves through the vocabulary of our body's gestures. When we are angry or sad, then we don't have to think in abstract terms of how we best manifest that state; rather, there exists a primordial rapport between our moods and bodies, such that the experience of being angry or sad is coexistent with our bodily existence.

The same is true of Shawn coming to a standstill. Something spooks him, a certain apprehension that is like that of an animal in danger. Anxiety then manifests itself in the very rigidity of the legs; the legs become ambassadors of anxiety. The expression of anxiety through the body is an invariant structure of agoraphobia more broadly. Time and again, the clinical literature points to vivid accounts of "symptoms" of anxiety being rooted in the rich vocabulary of the body: the body comes to a standstill, gesticulates wildly, and then sweats, trembles, before becoming weakened. The body stiffens and collapses, it acts irrationally and dissents from the image we have of it as being *one's own*. Such expressions of anxiety cannot be considered as isolated events to be managed and then removed. Rather, as Shawn's account makes clear, these symptoms only exist within relationship to a broader whole.

Not only is the body an expressive bearer of anxiety, it is also an object of anxiety. As Shawn comes to a standstill, so his body appears for him as a distinct thing. In turn, the thinglike quality of the body is itself an incitement to the production of further anxiety, as he writes: "you are very attuned to sensations in your legs, you will notice that they seem to have a mind of their own" (Shawn 2007, p. 119). Each of us knows from experience how our bodies can fail us or depart from us. At times, my body becomes an object for me against my own volition, such as when I am ill and feel my body as an impediment to my existence. On other occasions, I might experience a broader alienation from my body, such as when I see a photo of myself and fail to identify with the subject captured in the frame.

In these moments, we may well have an experience of the body as somehow distinct, other, or *thinglike* (Merleau-Ponty 1965, p. 209). That the body appears for me as different or even alien does not, of course, attest to substance dualism. Rather, the body's apparent distinction is maintained as a certain affective relation I have to my body. In general, these movements of self-alienation and bodily objectification are brief, and are often consolidated into a unified and strong sense of self, which accompanies us throughout the contingences and ambiguities of our perceptual existence. Agoraphobic anxiety is yet another way in which the body undermines the integrity of the self. In having a "mind of their own," the limbs do not appear as objects among many. Rather, their very thinglike quality is itself a dimension of anxiety, such that it is now "your very limbs [that] were demanding that you run" (Shawn 2007, p. 119). This shift from the body one possesses to the body that possesses one is instructive of the complex dynamic at work in phobic anxiety.

6 Anxiety as Spatial

We have seen how the body is inextricably tied up with anxiety in at least two respects: as expressive and objectlike. If we were to subtract these dimensions from our analysis, then we would be left with an impoverished sense of what it is like to suffer from agoraphobic anxiety. To complete this picture, however, two other aspects require attention: the spatial and intersubjective dimensions, each dimension underscoring the situated quality of emotion more generally.

Agoraphobia's spatial structure is framed by at least two major themes: spatial fragmentation and the centrality of home. Indeed, these two dimensions are intimately bound, such that if the agoraphobic person is able to move in the world, then it is thanks only to the construction of a rigidly established set of habits and patterns that establish an ongoing liaison with the home. By way of an illustration, consider several of the motifs appearing in Westphal's historic case studies: "He cannot visit the zoo in Charlottenburg, because there are no houses" (Knapp 1988, p. 60); "When in the company of a friend – he then experiences no fear of crossing spaces […] The crossing of spaces becomes easier when he stays next to a moving vehicle" (Knapp 1988, p. 66); "A cane or umbrella in his hand often makes the crossing easier" (Knapp 1988, p. 70). These examples reveal the highly situated and always conditional way in which people prone to agoraphobia move through the world. Lacking the freedom often taken for granted in "normal" instances of bodily existence, the subject has a tendency to rely on a proximity to familiar objects (the home), a means of escape (the car), or a prop employed to forge a spatiality of his or her own (the cane). In each case, the inevitable failure to maintain this tightly woven yet precarious grip on control leads to anxiety. The anxiety is no less present for Shawn. Problematic spaces tend to consist of either large expanses in which he is "trapped" in the middle, or narrow spaces flanked on all sides by vast space. Such spatial patterns are by no means uncommon, and the motif of homogenous or empty space is a recurring point of anxiety for many people prone to agoraphobia (Knapp 1988; Vincent 1919).

How is it that certain spaces gain a terrifying quality for agoraphobic people such as Allen Shawn? To answer this question let us turn back to his account. We note that spatiality is not a homogenous backdrop, upon which his anxiety takes place, as we find it in the clinical literature. Rather, as with bodily existence itself, spatiality becomes expressive of anxiety. In order for him to navigate the walk, Shawn must cut the space into manageable parts, as he has it: "I oriented myself by seeing how far I had walked from one small tree to another" (Shawn 2007, p. 117). Each tree in the landscape becomes a marker of how far or near he is to leaving the road. As such, Shawn's dissection of the road is reflective of the avoidance pattern of agoraphobes more generally to divide the world into safe and dangerous zones. To this extent, Shawn references Pascal as the archetypal agoraphobe. Pascal's terror in the face of the "eternal silence of infinite space" is duplicated on a microcosmic scale in Shawn's experience of upstate New York (Shawn 2007, p. 123).

Everything that opens up in homogenous space – vastness, impersonality, indifference, silence, abandonment – stands in radical contradistinction to the home.

Indeed, at the heart of the agoraphobe's anxiety is a concern of not being at home, both existentially and literally (cf. Trigg 2016). Consider here the function of the home within non-pathological experience. In everyday experience, we tend to carry a sense of being "at home" with us. We take this sense of being at home to mean a fluid interplay between ourselves and the world, such that we are able to move within the world without having to constantly interrogate where we have been and where we are going. Home is not a fixed site but is instead a certain bodily rapport we have with the world. In a word, in non-pathological instances of being-in-the-world there is generally a speaking a security that renders the home a diffused background presence – in the same way that the body is – rather than a focal point of our bodily and spatial existence.

Just as the body comes to the foreground for the agoraphobe as an obstacle in their existence, so the same is true of the home. For the agoraphobic person, home is not something carried with him or her as a tacit and spontaneous freedom in the world; rather, it stands out as a fixed and immovable site in the world, such that when this central source of orientation is removed, then the agoraphobe suffers from anxiety. In this light, we see, then, an empty road cannot be taken at surface level as a space objectively lacking danger. The "danger" is not manifest as a discernible feature of the landscape, but is instead marked as a symbolic potency, or as an expressive bearer of something beyond the road – that is to say, the non-home. Accordingly, faced with such a danger, Shawn feels that he might "disintegrate" (Shawn 2007, p. 123). Thematic experience is thus riddled with a series of sublevels that need to be attended to in order to understand how anxiety is both situated and far more complex than a disorder pertaining to one's non-situated "state of mind."

Here, the confluence between spatiality and embodiment comes into play in an especially striking way. To be sure, Shawn lists a series of symptoms that are common to experiences of agoraphobic anxiety, namely: "becoming short of breath and beginning to breathe rapidly […] feeling my heart beat at twice the normal rate […] finding my vision growing dark and blurred, feeling my face grow cold, and my legs tremulous, weak, and then extraordinarily stiff" (Shawn 2007, p. 117). But what prompts the expression of these symptoms is not simply a confrontation with open space, but instead – and critically – with disbelief that he will get to the end of this road and be the person he was when he started the journey. Far more than the production of merely uncomfortable sensations, the onset of symptoms suggests a more pervasive threat to Allen's self-presentation, at both pre-reflective and reflective levels. Indeed, the real issue is not the production of symptoms per se, but the threat those symptoms pose to the integrity of selfhood.

7 Anxiety as Intersubjective

We are forming a rich picture of the multifaceted nature of agoraphobic anxiety as a situated mood. As much as it is bodily, anxiety is also spatial. But agoraphobic anxiety is not an individual experience, such that other people assume a contingent or background role. Other people are constitutive of the experience itself either as a means of assuaging anxiety or through amplifying it. In the case of the former, the role of the "trusted person" assumes a defining importance in the subject's ability to traverse space without anxiety. Allen Shawn's own ability to cross the wide-open space is assisted not simply by the presence of "safety items" (Xanax, ginger ale, and a cell phone) but also by the presence of his companion, who "coaxes" him with the offer of a kiss as a "reward" (Shawn 2007, p. 118). In the same way that the car, umbrella, and proximity to home serve as "escape routes," so the same is true of the trusted person who accompanies the agoraphobic person in their anxiety. Their presence signals a familiarity, constancy, and understanding lacking in an otherwise precarious experience of the world (Trigg 2013a). Barlow describes a "safe person" in the following respect: "A safe person is commonly a significant other whose company enables the patient to feel more comfortable going places than he or she can be either alone or with other people. Usually, this person is considered 'safe' because he or she knows about the panic attacks" (Barlow 2002, 343). Having knowledge of the patient's panic attacks not only disarms the efficacy of the panic attack but also provides a legitimate context to manage anxiety should the subject be "incapacitated by panic" (Barlow 2002, p. 343). As a result, in the company of the trusted other, the agoraphobic person is able to maintain a stronger sense of identity than if he or she were alone.

The function of the so-called "trusted other" has been documented extensively, but let me cite one illustrative case from Westphal concerning a patient named Mr. C.

> The same feeling of fear overtakes him when he needs to walk along walls and extended buildings or through streets on Holiday Sundays, or evenings and nights when the shops are closed. In the latter part of the evening – he usually dines in restaurants – he helps himself in a peculiar way in Berlin; he either waits until another person walks in the direction of his house and follows him closely, or he acquaints himself with a lady of the evening, begins to talk with her, and takes her along until another similar opportunity arises, thus gradually reaching his residence. Even the red lanterns of the taverns serve him as support; as soon as he see one his fear disappears (Knapp 1988, pp. 60–61)

Here, we see how anxiety becomes mediated by the presence and absence of other people. In the presence of others – even if they are strangers – anxiety is diminished thanks to the fact the other person signifies a reliable presence that the agoraphobe himself is lacking. Other people present themselves as steadying and constant, and they are depended upon in the sense not only of being present in case something goes wrong with the agoraphobe – a heart attack, a fainting spell – but also in terms of being a beacon of familiarity in an otherwise unpredictable and indifferent world. There is thus a case to be made for both the situated quality of emotion as well as its extensionality. Anxiety is neither in one's head nor even in one's body. Rather, it forms an arc in and around a subject, and affects everything within its reach (Trigg 2013b). In the anecdotal evidence, time and again we see how the trusted companion of the agoraphobe becomes enmeshed in the spatial and material experience of being-in-the-world. Here is a report from 1884, not long after the term agoraphobia was coined. Of one agoraphobic subject, we read how:

> Companionship relieves the feeling of loneliness and fear produced by the thought of taking a holiday in a part of the country new to him. The presence of a cart, even a stick or umbrella in the hand, persons, or trees, gives a sense of confidence when walking an unknown road. Cheerful and lively conversation, with a congenial companion, will always ward off the attacks (White 1884, p. 1140).

In another report, we are told how:

> Many totally avoid being alone, while others require a companion only when venturing beyond their '"safety zones." The comfort provided by the companion may be complete; that is, the agoraphobic person's activities are unrestricted while in the companion's presence .... Some agoraphobic's require the presence of one particular person, usually a family member, but others are less selective and may be reassured by even the company of a pet (Chambless and Goldstein 1982, p. 2).

In each case, the affective experience of space is augmented by the presence of other people. Other people serve not only to modify the relation the subject has with his or her anxiety; their presence also transforms the felt experience of the world itself. In the presence of the trusted other, the world opens itself, such that the agoraphobe is able to move with far greater freedom than were he alone. Such a modification does not derive from an analytical reflection on the factual presence of another person. As we see above, the role of a trusted person may also be assumed by an inanimate prop. What generates the experience of comfort stems from a prepersonal situatedness that brings together embodiment, spatiality, and interspatiality in one coherent frame, such that we cannot consider one of these dimensions without in turn considering the other.

Other people do not always assuage the experience of anxiety; they can also amplify and reinforce an already existing anxiety. Clinical research on the role of other people in the development of agoraphobic anxiety suggests that the look of other people is a significant factor in precipitating the onset of panic (cf. Davidson 2000a, 2000b, 2003). In fact, a heightened self-consciousness regarding how other people perceive the subject is consistent with the desire to maintain the self-presentation of being a "normal" and "healthy" individual both to oneself and to others (Vincent 1919). In this respect, other people are a critical problem for many agoraphobic people.

Whereas spatial routes and bodily habits can be controlled to some extent by developing a set of habitual patterns that render perceptual experience predictable, exerting control over how other people perceive us remains less likely. To this end, the very centrality of the home as the safe place *par excellence* is predicated on its function as concealing the look of the other, as Joyce Davidson notes, "[s]ufferers' homes are frequently organized to minimize the fear of the look" (Davidson 2003, p. 84). Unlike inanimate props such as cars and umbrellas, other people are not simply objects for our own use, but also perceive us as objects in the world. As objectified by the look of the other, the attempt at maintaining a presentation of being "normal" for the subject proves contentious. Through the look of the other, the attempt at concealing anxiety through adhering to a ritualized and regulated life risks being detected, and in being detected, the very anxiety that the subject seeks to mask from the world in turn becomes an object of interrogation for the other person.

8 Conclusion

Our aim in this chapter has been to demonstrate how anxiety is a situated affect. Whereas thinkers such as Heidegger have given rich conceptual accounts of anxiety, his account nevertheless remains abstract in terms of omitting the bodily, spatial, and intersubjective dimensions. By contrast, as we have seen, Heidegger frames anxiety primarily as a transformative state, able to instigate a profound shift in our being-in-the-world. The same is true generally speaking of the clinical treatment of anxiety. A third-person perspective on anxiety tends to treat the body in objective terms as a set of processes to measure and observe. The lived and situated quality remains lost, however, in this approach. In the final case, the treatment of conditions such as agoraphobia reinforces the understanding of anxiety as a state of mind, the management of which requires a change of thinking.

In contrast, we have sought to show that prior to its conceptualization as a transformative state, agoraphobic anxiety is richly construed as a situated experience. This has been evident in at least three respects. In the first case, agoraphobic anxiety is revealed in and through the lived body, where the body is both expressive of anxiety and also an object of anxiety. In the second case, spatiality is not a passive backdrop against which agoraphobia takes place, but is instead its medium of expression – manifest here as a fragmented spatiality, which revolves at all times around the presence and absence of home. In the final case, other people are neither bystanders in the onset of agoraphobic anxiety nor contingent to its development. Rather, other people both assuage and amplify anxiety in equal measure. Indeed, it is a testament to the situated quality of anxiety that an agoraphobic person is both inhibited and liberated by and from their anxiety depending on their intersubjective relations.

There is, of course, much more to say on the situated quality of anxiety, and in the present chapter we have confided ourselves solely to agoraphobic anxiety. We have done this because agoraphobia is an especially striking case of how emotion can only be understood by taking the condition in a relational whole. In cases of generalized anxiety, the analysis would lend itself to a more diffused though no less present account of situatedness. Likewise, there is much more to say on how the collective status of anxiety – an atmosphere of anxiety – serves to contest the notion of anxiety as an experience undergone by individuals alone (one can think here of the current climate of anxiety concerning borders and boundaries, and how this climate carries with it an atmospheric quality, seeping as it does from one medium to another). Such developments, however, require a separate treatment of the complex situated dimension of anxiety.

References

Aslam, N. 2012. Management of panic anxiety with agoraphobia by using cognitive behavior therapy. *Indian Journal of Psychological Medicine* 34 (1): 79–81.

Barlow, D. 2002. *Anxiety and Its Disorders*. New York: Guilford Press.

Chambless, D., and A. Goldstein. 1982. *Agoraphobia: Multiple Perspectives on Theory and Treatment*. New York: Wiley.

Davidson, J. 2000a. …the world was getting smaller': Women, agoraphobia and bodily boundaries. *Area* 32 (1): 31–40.

———. 2000b. A phenomenology of fear: Merleau-Ponty and agoraphobic life-worlds. *Sociology of Health and Illness* 22 (5): 640–660.

———. 2003. 'Putting on a face': Sartre, Goffman and Agoraphobic Anxiety in Social Space. *Environment and Planning D: Society and Space* 21 (1): 107–122.

Edelman, R.E., and D. Chambless. 1993. Compliance during sessions and homework in exposure-based treatment of agoraphobia. *Behaviour Research and Therapy* 31: 767–773.

Heidegger, M. 1977. *Basic Writings*. Trans. David Farrell Krell. New York: Harper Collins.

———. 2008. *Being and Time*. Trans. J. Macquarrie and E. Robinson. New York: Harper Perennial Modern Classics.

Knapp, T. 1988. *Westphal's 'Die Agoraphobie' with Commentary: The Beginnings of Agoraphobia*. Trans. T. Michael Schumacher. Lanham: University Press of America.

Martin, R., and E.R. Dahlen. 2005. Cognitive emotion regulation in the prediction of depression, anxiety, stress, and anger. *Personality and Individual Differences* 39 (7): 1249–1260.

May, Rollo. 1996. *Meaning of Anxiety*. New York: Norton and Company.

Merleau-Ponty, M. 1965. *Structure of Behavior*. Trans. A. Fisher. Boston: Beacon Press.

———. 2012. Phenomenology of Perception. Trans. D. Landes. New York: Routledge.

Prosser White, R. 1884. Agoraphobia. *The Lancet* 124: 1140–1141.

Shawn, A. 2007. *Wish I Could be There: Notes from a Phobic Life*. New York: Viking Press.

Trigg, D. 2013a. The body of the other: Intercorporeality and the phenomenology of agoraphobia. *Continental Philosophy Review* 46 (3): 413–429.

———. 2013b. Bodily moods and unhomely environments: The hermeneutics of agoraphobia and the Spirit of place. In *Interpreting Nature: The Emerging Field of Environmental Hermeneutics*, ed. F. Clingerman et al., 160–177. New York: Fordham University Press.

———. 2016. *Topophobia: A Phenomenology of Anxiety*. London: Bloomsbury.

Trigg, D., and S. Gallagher. 2016. Agency and anxiety: Loss of control and delusions of control in schizophrenia and anxiety. *Frontiers in Human Neuroscience* 10: 459.

Vidler, A. 2000. *Warped Space: Art Architecture, and Anxiety in Modern Culture*. Cambridge, MA: MIT Press.

Vincent. 1919. Confessions of an Agoraphobic Victim. *The American Journal of Psychology* 30 (3): 295–299.

CPSIA information can be obtained
at www.ICGtesting.com
Printed in the USA
BVHW040226081118
532517BV00005B/18/P

9 783319 929361